T0305965

EMI Filter Design

THIRD EDITION

EMI Filter Design

THIRD EDITION

Richard Lee Ozenbaugh   Timothy M. Pullen

CRC Press is an imprint of the
Taylor & Francis Group, an informa business

CRC Press
Taylor & Francis Group
6000 Broken Sound Parkway NW, Suite 300
Boca Raton, FL 33487-2742

First issued in paperback 2017

Version Date: 2011906

ISBN 13: 978-1-4398-4475-5 (hbk)
ISBN 13: 978-1-138-07407-1 (pbk)

Library of Congress Cataloging-in-Publication Data

Ozenbaugh, Richard Lee.
EMI filter design / Richard Lee Ozenbaugh, Timothy M. Pullen. -- 3rd ed.
p. cm.
Includes bibliographical references and index.
ISBN 978-1-4398-4475-5 (hardcover : alk. paper)
1. Electric filters--Design and construction. 2. Electromagnetic interference. I. Pullen, Timothy M. II. Title.

TK7872.F5O93 2012
621.3815'324--dc23 2011034035

Visit the Taylor & Francis Web site at
http://www.taylorandfrancis.com

and the CRC Press Web site at
http://www.crcpress.com

Contents

Preface

In today's world of Internet TV, 4G cell phones, the ever-growing list of exotic coffee blends, and the iPad, who would think for even a moment that it would still be necessary to add large, strangely shaped components into equipment to ensure that EMC regulations are met? Has time stood still for EMI? Perhaps software should be providing this functionality for us today? Unfortunately, that's not going to happen. EMI filters offer a means of protection between the outside world and the inner workings of our equipment. These filters are most often placed at the inputs of the equipment in order to restrict conducted noise emissions propagating to external electrical networks, which could damage or interfere with other electronic equipment connected to the same electrical source. Furthermore, these filters also restrict noise on the very same external power lines from entering the equipment. In principle, this may sound simple enough; however, filters are not quite so simple to design. One thing we can be sure of, EMI filters are not likely to disappear in the very near future and are a very important aspect of any system.

EMI filtering is a necessary evil, and with power systems becoming more efficient as switching frequencies move into the 500-kHz and beyond range, the need for robust EMI filters is even more essential to ensure EMC compliance for both conducted and radiated emissions. Moreover, these last few years have seen continued growth in the use of motor control within the realms of the more-electric aircraft (MEA). Removing heavy and cumbersome hydraulic systems and replacing them with advanced motor control platforms has posed a significant challenge to engineers in terms of weight, reliability, thermal compatibility, and EMC. This is a high-growth aspect of power electronics and power system management where strict EMC requirements must be met. EMI filter design is often treated with contempt, very much like power supplies are sometimes the last piece of the puzzle to be considered in a larger system. "It's something we have used before"; "I am sure we can get the power we need from the supply we used last year"; and "The EMI filter that we used on a previous program seemed to pass okay." Do these sound familiar? In today's world where systems are required to offer increased levels of performance and reliability, such as in the aerospace electronics industry, we find that cost, size, weight, performance, etc., are all critical-needs factors that must drive the design from the outset. EMC and EMI filter design is also a critical aspect of the system that deserves strict attention early on in the concept design stage. In defining architectures and performing systems analysis, valuable questions can be asked, such as

- What is the input power architecture?
  Three-phase/single-phase AC
  Single ended (grounded)
  Differential (floating)
- What is load power conversion topology?
- How much EMI filter volume is needed for a specific power conversion topology?
- What type of filter is needed to realize the insertion loss required?
- Can the filter be optimized to reduce the effects of higher switching currents?
- What is the impact of EMC failure?
- Is recovery from EMC failure possible without a redesign?

These are valid questions that can be applied to almost any EMI filter solution for any equipment type and in any industry. Needless to say, mission-critical applications that demand high performance require robust solutions based upon formal analysis and not just trial and error. EMI filters can, to some degree, be designed using simulation and analysis, and most certainly for differential-mode loss. This is based upon the accurate assessment of the harmonic content and amplitude, and in developing a filter solution that provides sufficient insertion loss over the limit frequency range. If the filter is designed using this approach, then EMC testing has a very good chance of being successful. To a larger degree, we may also suggest that radiated emissions should be relatively low if conducted emissions are well within specification. Common-mode interference is sometimes inherent in a system design, but most often it is inductively or capacitively coupled from an external source. In a motor drive, for example, the switching action and localized parasitic capacitance will create common-mode interference. To be able to determine these levels is not as simple as in the case of differential-mode harmonic analysis; therefore, preliminary testing and defensive design are good risk mitigation strategies. In many cases, common-mode rejection is also mitigated by careful design of switching circuits, controlling parasitic influence, careful control of switching rise and fall times, etc. EMI filter design is often a complex aspect of the overall system and needs to be carefully engineered, not overdesigned or left to chance. On the flip side, EMI filter design is not a precise science in terms of accurate placement of poles, or of meeting a very accurate –3-dB pole-Q frequency; it's about placing insertion loss over a range of frequency bands to reduce conducted emissions to acceptable levels.

The second edition of this book was written by Richard Ozenbaugh and published in 2000. This third edition is a consolidation of topics from the second edition while also presenting new material that covers some of the analysis techniques necessary for passive filter realization. The text also discusses the approaches for LC filter structure design and includes a more practical hands-on look at EMI filters and the overall design process. This third edition is also a collaborative effort and has been written as a book for EMI beginners and those who are interested in the subject. There is no hard-and-fast definitive solution to EMI filter design, and there are obviously many concepts and potential solutions to a unique problem. This book presents a methodology for design that is used by the authors, and it is hoped that this text will cultivate new ideas and more effective solutions by many of the readers. As always, the responsibility for this book remains with the authors alone. We hope you find it useful.

Acknowledgments

I wish to thank Colonel W. T. McLyman, NASA Jet Propulsion Laboratory, now retired, for his KGMAG computer programs, his technical help, and the transparencies; Armando Valdez, Aerojet General, Azusa, California, for all his recommendations and consulting; Bob Rudich, Airesearch, Torrance, California, for another design approach; Robert Hassett, now retired, RFI Corp., Long Island, New York, along with the rest of the RFI engineering group, for all their support, patience, graphs, suggestions, papers, and critiques; Mitchell Popick, for all his help, suggestions, and argumentation, which often forced me to think in different terms; the people at Powerlab, Pomona, California, for their backing and help, primarily with the power supply equations; the people of the James Gerry Co., especially Jim Gerry, who helped set up the seminars that started this whole thing; Harvey Gramm, URAD, for all the special computer time, ideas, transparencies, and the printing; my son, Richard Lee Ozenbaugh II, for the help with the EMI Design computer program development. Finally, I pay special thanks to my wife, Pansy, for allowing me to spend such a long time hidden away in my room writing these three editions.

Richard Lee Ozenbaugh

Thanks to "Oz" (Richard Ozenbaugh) for his friendship and for allowing me to contribute to the third edition; thanks also to my friends and peers within HDDC, all of whom have read the second edition of this title, for their feedback and ideas. I would also like to mention a few people who have, in one form or another, influenced my career over the years: my mother and father, Susan and Geoffrey; Andrew Smolen, Richard Clark, Michael C. Hulin, Dr. Ron Malyan, and Dr. Hari Bali. Finally, special thanks go to my wife, Linda Marie, for her quiet encouragement.

Timothy M. Pullen

Authors

Richard Lee Ozenbaugh has been a contributor within the electrical and electronic arena since the early 1950s. Richard states, "I have designed it, built it, taught it and sold it." In his early days, he was a veteran of the U.S. Navy as a radar specialist. Ozenbaugh attended the University of Nebraska and received his engineering degree from the Capitol Radio Engineering Institute in Washington, D.C. He started working in magnetic houses during the 1970s, which included EMI filter design. He joined Hopkins Engineering in the 1980s and later moved to RFI on Long Island. He spent many years as a consultant within the fields of magnetic component design and EMI filter design. Ozenbaugh wrote the first edition of *EMI Filter Design* in the 1990s and the second edition in 2000. Among the many years spent developing EMI filter solutions, Ozenbaugh has presented various seminars at the professional level to many key companies such as Hughes Aircraft Corporation, Smiths Aerospace, Parker Hannifin Aerospace, Franklin Electric, McDonnell Douglas, Breeze Eastern, Cirrus Logic, and many others. He has also consulted with a number of these companies.

Timothy M. Pullen is a principal electrical engineer with Rockwell Collins. Originally from the United Kingdom, and a graduate in electrical and electronic engineering, Pullen has over 25 years experience in research, design, and development of electronic systems for both commercial and military applications, including power electronics, motor control, and FADEC (full authority digital electronic control) technology. His career includes working for British Aerospace (U.K.), Smiths Industries Aerospace (U.K.), Honeywell (U.S.), and Parker Hannifin Aerospace (U.S.). In his current role within HDDC Power Electronics group, Pullen is a key contributor on several large programs, including BLDC (brushless DC) and ACIM (AC induction motor) motor controller development, where he is responsible for the architecture development and design of control systems used for flight-surface actuation. His expertise includes model-based simulation and dynamic modeling of control systems, power electronics, motor control, low-noise analog circuit design, and EMI filter design.

Terms and Abbreviations

AC: alternating current
ADC: analog-to-digital converter
BH: flux density and hysteresis curve for magnetic core materials
C, C1, C5: capacitor—numbered and unnumbered
CE: conducted emissions
CE-101, CE-102: MIL-STD-461 conducted emissions requirements
CIP: current-injection probe
CISPR: Comité International Spécial des Perturbations Radioélectriques
CM: common mode
DAC: digital-to-analog converter
dBμA: decibel (dB) microampere
dBμV: decibel (dB) microvolt
DC: direct current
DM: differential mode (normal mode)
EMC: electromagnetic compatibility
EMI: electromagnetic interference
FCC: Federal Communications Commission
FFT: fast Fourier transform
FSLM: frequency-selective level meter
GHz: gigahertz (billion cycles per second)
HEMP: high-energy magnetic pulse
HF: high frequency
HIRF: high-intensity radiated field
Hz: hertz (cycles per second)
kHz: kilohertz (kilocycles per second)
L: two-component filter network shaped like the letter L
L, L3, L9: inductor—numbered and unnumbered
LISN: line-impedance stabilization network
MHz: megahertz (megacycles per second)
MOSFET: metal oxide semiconductor field-effect transistor

MOV: metal oxide varistor
MPP: molybdenum permalloy powder
mu (μ): Greek letter μ (permeability)
PCB: printed circuit board
pi (π) filter: three-component filter network shaped like the Greek letter π
PSD: power spectral density
PWM: pulse-width modulation
Q: Q-factor or quality factor of LC filter
R, R5, R10: resistor—numbered and unnumbered
RE: radiated emissions
SNR: signal-to-noise ratio
SRF: self-resonant frequency
T filter: three-component filter network shaped as T section
TVS: transient voltage suppressor; Transzorb
y, y_{11}, y_{22}: admittance—the reciprocal of impedance
z, z_1, z_0: impedance
zeta (ζ): damping coefficient of complex pole pair
Zorro: common-mode inductor

Filter Types

AP: all pass
BP: band pass
BR: band reject
HP: high pass
LP: low pass

Organization of the Book

The scope of this book is EMI filter design and the practical application of formal techniques that will enable the reader to develop simple filter solutions. The book is partitioned into 20 chapters.

Chapters 1 and 2 provide an introduction to the book. Chapter 3 looks at the causes of both common- and differential-mode noise and methods of elimination. Chapter 4 discusses the source and load impedances for various types of input power interfaces. Chapter 5 looks at the load impedance aspect of EMI filter design. The next six chapters cover EMI filter structures and topologies and provide discussion on components. Chapter 12 discusses voltage transients and provides insight into sizing of components and protection. Chapter 13 looks at issues that will compromise filter performance. Chapter 14 presents a summary of the types of noise seen by both equipment and EMI filters. The chapter also provides their mathematical equivalence in terms of Fourier representation. Chapter 15 discusses filter requirements and presents a design goal for a filter design objective. Chapters 16 and 17 present a matrix method of design of filters using matrices. Chapter 18 presents two-port analysis and explains the transfer function method of LC structures and their equivalent polynomials. Chapter 19 provides the reader with a design example, including a discussion of circuit application and analysis techniques. Chapter 20 presents packaging solutions of EMI filters.

1 EMI Filters

1.1 Introduction

Conceptually, an electrical filter network will filter out lower or higher frequency bands while passing specific bands of frequencies. The basic property of an EMI filter is usually described by the insertion-loss characteristic. That characteristic is typically frequency dependent, and it refers to the attenuation of the EMI filter. The measurement of insertion loss is complicated by several aspects. The configuration of the input and output terminals of an EMI filter is changed through different types of measurement setups, and this fact alone complicates the measurement itself.

Another problem is represented by nondefined impedance terminations at the input and output sides of the filter. The impedance of the power supply network is connected to the input terminals of the EMI filter. The current impedance value of the power supply network depends on the type of the power network, current load, and also on the operating frequency of the test signal. The output of the filter is generally loaded with impedance, which is usually unknown and not steady in the time domain.

EMI filter design engineers think in terms of attenuation, insertion loss, and filter impedance, while a regular wave filter designer thinks in terms of poles, zeros, group delay, predistortion, attenuation, and the order of the filter. In both cases, the concepts are mathematically the same; however, EMI filtering is not a precise science as is, for example, an active low-pass filter that might be used as an anti-alias filter within a data-acquisition application. In this particular case, accurate placement of the –3-dB corner frequencies is essential, as the filter must protect the ADC from HF-folded spectral components.

EMI filters are all about presenting a high impedance to a given range of frequencies to provide sufficient insertion loss at those frequencies. This suggests that the role of an EMI filter is to create maximum mismatch impedance at undesired frequencies while providing maximum matching impedance at the desired frequencies so that they pass the filter unchanged. Accurate placement of so much dB loss at a particular frequency is not always necessary as long as the filter provides the insertion loss where it is required. Insertion loss is simply the ratio of the signal level (v_1) in a test configuration without the filter installed relative to the signal level (v_2) with the filter installed. This ratio is typically described as follows:

$$IL(dB) = 20\log_{10}\frac{|v_1|}{|v_2|} \tag{1.1}$$

The presence of both differential- and common-mode insertion loss will lead to attenuation of both differential- and common-mode interference artifacts, respectively.

To be effective, higher levels of insertion loss will lead to increased levels of attenuation. The EMI filter is an all-pole network where series elements (inductance) have high impedances, while the shunt elements (capacitance) offer low impedance to the unwanted frequencies.

Figure 1.1 represents a typical EMI filter structure that could be used in single-phase AC applications or for DC power inputs. The filter provides both common- and differential-mode loss along with two dQ RC shunt networks (R1,C3 and R2,C7). In PWM switching converters, incremental negative resistance will force a maximum filter output impedance to ensure stability. These compensation shunt networks provide a means of controlling the filter Q and modifying the output impedance. L1 is a common-mode choke; L2 and L3 form the differential-mode inductance. Line-to-line capacitors C1, C2, and C4 are used for differential-mode loss. C5 and C6 are used in conjunction with L1 and provide second-order loss to common-mode noise artifacts. This filter structure is further discussed and analyzed in terms of design and performance in chapter 19.

Designing an EMI filter to meet a unique EMC requirement is often a challenging exercise, and simply defining a –3-dB corner frequency and expecting the filter to do the rest is not always going to bring a successful conclusion to the EMC solution. EMI is not always deterministic, especially in the case of high-power applications where PWM is present. Meeting both differential- and common-mode loss requirements very often requires up-front analysis of the power electronics architecture, knowledge of PWM topologies and switching frequencies, etc., while also making use of circuit synthesis and

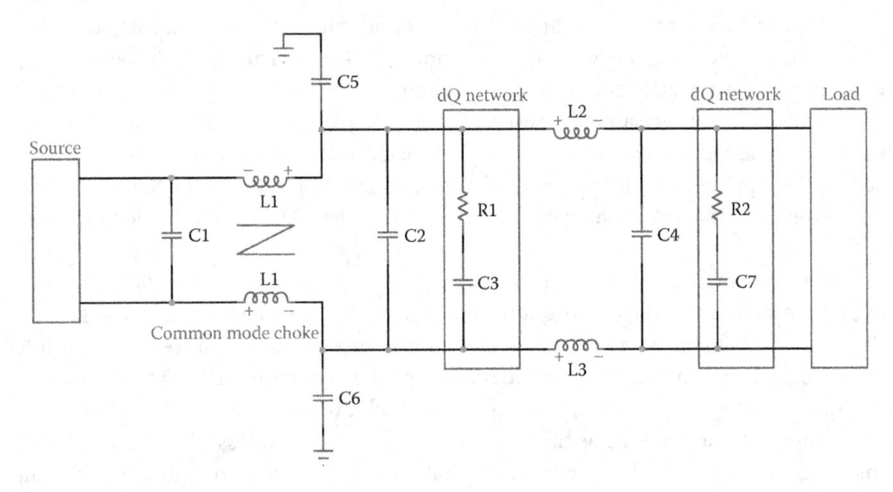

FIGURE 1.1 EMI filter circuit structure.

simulation. With all this data in hand, defensive design along with experience will make it possible to define a filter that has a good chance of meeting unique EMC requirements such as DO-160 and MIL-STD-461.

1.2 Technical Challenges

EMI filter design is not a clear-cut case of applying a set of poles to a textbook polynomial, nor is it a simple case of applying a corner frequency to an LC circuit structure with the expectation that the frequency-magnitude slope or insertion loss versus frequency will ensure EMC success. EMC describes a state in which the electromagnetic environments produced by natural phenomena and by other electrical and electronic devices do not cause interference in electronic equipment and systems of interest. Of course, to reach this state, it is necessary to reduce the emissions from sources that are controllable, or to increase the immunity of equipment that may be affected, or to do both of these. In mission-critical applications, doing both is absolutely essential to ensure safety, robustness, and reliability.

It is important to understand that EMC as defined does not absolutely prevent interference from occurring. Emissions from various sources are variable; lightning impulses on power and signal lines, for example, vary with the level of lightning current and its proximity from the equipment. In addition, the immunity of a particular piece of equipment can vary; exemplifying this case is the fact that induced voltages on a circuit board are strong functions of the angle of incidence and the polarization of the incident electromagnetic field. Recognition of this variability will ensure finding a balance between immunity and emissions for a particular type of disturbance, which should be sufficient to prevent EMC problems in most cases.

EMC is also dictated from a design standpoint and in the selection of a particular power conversion topology. In almost all cases today, these are PWM converters. Within switch-mode power supplies, a DC voltage is switched at a high frequency that can range from the low kilohertz to 200 kHz and beyond. This high-speed switching process is intrinsic to switch-mode power supplies, and it provides greater efficiency and reduced size than linear power supplies. However, as a side effect, this switching generates unwanted EMI. In fact, most conducted EMI within switch-mode power supplies originates from the main switching MOSFETs, transistors, and output rectifiers. In either power supplies or electronic equipment, it is the function of the EMI filter to keep any internally generated noise contained within the device and to prevent any external AC line noise from entering the device. Because unwanted EMI is at much higher frequencies than normal signals, the EMI filter works by selectively blocking or shunting unwanted higher frequencies. The inductive part of the EMI filter is designed to act as a low-frequency pass device for the AC line frequencies and as a high-frequency blocking device. Other parts of the EMI filter use capacitors to bypass or shunt unwanted high-frequency noise away from the sensitive circuits.

The net result is that the EMI filter significantly reduces or attenuates any unwanted noise signals from entering or leaving the protected electronic device. Certain applications such as high-power switch-mode supplies yield very high common-mode currents due to PWM switching, and these typically will be seen at higher frequencies. Capacitive coupling between power switches and chassis will provide paths for these currents, and in some cases, the estimated common-mode insertion loss within the EMI filter will be too small. This, therefore, will lead to noise artifacts at higher frequencies, and if the filter is not optimized for characteristic impedance or, as in the case of cascade LC structures, the filter sections are not optimized for adjacent input-output impedance, the filter will be prone to peaking at the points of impedance mismatch. If the EMI filter design is left unchecked, this phenomenon may lead to corresponding elevations in noise amplitude around the pole-Q frequency, causing EMC test failure. Overall, the challenges of EMI filter design are more prevalent in higher power applications where PWM is employed.

1.2.1 Controlling Parasitic Uncertainty

Along with all the technical challenges that surround EMC and the design of an optimal EMI filter, there are other issues that must be carefully considered, and many of these are under the direct control of the engineer. Care must be taken with the physical layout of the filter. This should be undertaken not just for the passband frequencies, but more importantly for the frequencies in the stop band that may be well in excess of the cutoff frequency of the low-pass filter. Capacitive and inductive coupling are the main elements that cause the filter performance to be degraded. Accordingly, the input and output of the filter should be kept apart. Short leads and tracks should be used, and components from adjacent filter sections should be spaced apart. It is often the case that an EMI filter is placed directly onto a PCB adjacent to other sensitive circuits. Physical routing of power cables to and from the PCB, including traces within the EMI filter structure are, in some cases, not optimized for low impedance and impedance balancing. With these design and packaging limitations, EMC failure may very well be attributed to the layout and packaging of the filter.

1.3 Types of EMI Filters

1.3.1 AC Filters

EMI suppression filters for AC power lines eliminate noise entering equipment from commercial power lines or noise generated from electronic equipment that may be sharing the same AC power connection. In general, a single-phase AC filter is topologically the same in circuit structure as its DC filter counterpart. Design of the AC filter must look at the selection of both the differential mode (X) and common mode (Y) capacitors, ensuring that they are rated correctly for the peak voltages of the application. Typically, common-mode chokes along with X and Y capacitors are generally used for AC EMI noise suppression for the filter. In suppressing common-mode noise, common-mode chokes are the most important components because their characteristics influence the overall performance of the filter.

There are two types of common-mode chokes: the standard type and high-frequency type. A hybrid choke coil is a high-performance EMI suppression choke that can suppress both common-mode noise and differential-mode noise at the same time. It is effective in AC power supplies where a need for higher frequency harmonic countermeasures is essential. Another type of AC filter is used on three-phase power systems. This filter is often a balanced common-mode core, but it can also include additional line-to-line differential-mode loss.

1.3.2 DC Filters

EMI filters are typically used on equipment that is DC sourced, such as DC-DC power converters, motor controls, pump controls, etc., and includes a significant number of aircraft systems. The DC filter will look the same as a single-phase AC filter, but it will not need to be voltage derated to the same levels for both the X and Y capacitors. Being DC, the return path for the current can either be fully differential to the point of source or grounded at the equipment. The latter has huge ramifications to a potential EMI design solution, as common mode must be based upon a differential power interface. If a DC-fed system has its ground connected to chassis, then only the live DC input feed needs to be filtered. Thus, the filter need only be single ended in topology. Usually, DC filters are interfaced to a DC link that has a bulk capacitor or connects to the input of a PWM power converter. This also drives the need for a careful understanding of the system, as the filter output impedance must be defined so as not to interact with PWM systems that have incremental negative input impedance. If the filter impedance is equal to or greater than the power converter negative input impedance, oscillation is a certainty. To counter this, it is often necessary for the filter to have a shunt RC network that will modify the filter output impedance so that oscillation does not occur. This is discussed in detail in chapters 18 and 19.

1.4 No Such Thing as Black Magic

EMI is often considered *black magic*, when it is really nothing of the sort. EMI relies on electromagnetic-wave and transmission-line theory and comes about through second- or third-order effects. Some of these effects exist without an engineer's knowledge, most often through poor system design or a lack of EMC awareness. This is one of the reasons why some people call it black magic.

In many cases, when we talk about EMI, we are relating to two components, namely electric (E-field) and magnetic (H-field). Both of these run perpendicular to each other. Therefore, we may conclude that EMI is a function of current, loop area, and frequency. These subject areas are often neglected or left to chance and are one of many reasons why EMI is misunderstood by many engineers who are not familiar with the subject. An EMI filter, if not designed correctly, can actually make EMC impossible to manage.

There are so many variables to consider with EMI design that it is impossible to define a precise mathematical solution. The EMI circuit structure solution is complex and must be based upon knowledge of the complete system, the source, the power conversion topology, the quality of the PCB layout, and the grounding structure. These all play against each other and, in reality, there is never an optimal solution. Meeting an EMI specification is all about balancing system immunity against emissions so that the requirements are met. There are no strict rules and no definitive mathematical solutions that can be applied to define a precise circuit solution. Because of this variability, each case is unique and requires its own solution. As we noted previously, balancing immunity against emissions is all one can do.

1.5 It Is All in the Mathematics

EMI filter design can be accomplished in a rule of thumb fashion where estimation of losses, pole-Q frequency, and stability are all based upon previous experience and guesswork. Needless to say, unless the power system or equipment is familiar to the filter designer, it is likely that the filter will need to be tuned considerably, which points to another statement. EMI filters need to be designed with some degree of flexibility so that they can be modified during test configuration. This will allow the engineers to make changes to both common and differential pole-Q frequencies while ensuring filter stability in any configuration should the need arise.

The question of EMI filter design and the use mathematics is posed. We know how to compute the pole-Q frequency of a double-pole LC filter. So, what differential-mode insertion loss do we need? Where do we place the pole-Q frequency? How many LC stages do we need? How do we stabilize the filter? Does the filter output impedance ensure stability when connected to the input of a PWM converter that has inherent incremental negative impedance? What about common-mode insertion loss?

There are many factors that help to drive a successful EMI filter solution, and some of these are actually unknown, which adds to the mystique and "black magic." We know now that black magic is really another name for *parametric uncertainty.* These are the factors and inherent physics that are not realizable from the outset. We could suggest common-mode leakage, parasitic elements, source and load impedances, etc., to name a few. And so, what do we do with the mathematics? Mathematics will allow the filter solution to be numerically approximated through a series of iterative steps so that a ballpark circuit solution can be achieved. These methods include matrix analysis, polynomial expansion of transfer functions, circuit simulation, etc. These are all presented in detail within chapters 16–19.

We must remember that to get from the ballpark solution to seat 24 in row G will take fine-tuning and modification of the test configuration. Therefore, the filter must be designed defensively.

2

Why Call EMI Filters Black Magic?

In some cases, engineers—both designers of electromagnetic interference (EMI) devices and others—call EMI black magic. As we have already said in chapter 1, it is nothing of the sort. There are, however, several main reasons for this misconception. First, there is no definitive mathematic solution or well-defined design method that can be applied to all EMI filters due to the unique application in each case and the variability in both source and load impedance over the frequency bands of interest. These two factors provide significant challenge, even more so if the filter is required to meet very stringent requirements such as MIL-STD-461 for CE101 and CE102.

Another reason why EMI filter design has some level of mystique attached to it is probably due to the complexity of the design process along with the various variables that all drive a successful design. To name a few, these could very well include parasitic uncertainty, insertion loss, common mode versus differential mode, defining the PWM current signature, FFT versus PSD, impedance matching of adjacent LC structures, filter stability, filter characteristic impedance, filter output impedance versus load impedance, etc. Typically, unless one is very familiar with passive network theory, either through experience or filter design of two-port networks, etc., these subjects are either a foreign language or were left at the university a long time ago. EMI filter design is often a small piece of the pie, and this is certainly so for complex power systems. From this perspective, EMI filter design might appear to be simplistic, but in reality it is a complex subject and requires a solid understanding not just of circuit network theory, but also of the system that it is being designed to protect.

For the purpose of clarity, if we are presented with an EMI requirements specification, we know what we are connecting to in terms of voltage, current, source, and load and, therefore, are able to approximate impedances. Furthermore, if we have a good approximation of the differential-mode current signature in the case of a PWM-based power converter, then we are able to create a defensive EMI solution that would potentially meet the needs of the filter for both differential-mode and common-mode loss while also managing filter stability. We say *defensive*, as the filter will very likely need tuning during testing. So, additional nonpopulated components added for X and Y capacitors, including damping dQ networks, may well prove to be successful. No black magic or wizardry is involved: just good old-fashioned engineering with a blend of mathematics, physics, and electromagnetic field theory.

2.1 What Is EMI?

EMI is electromagnetic interference. It is also called conducted emissions or radiated emissions. This book covers mainly conducted emissions (CEs), which means any unwanted signal or noise on the wiring or copper conductors. The reason for the reference to power cabling is that EMI filters are part of the power wiring and are designed to remove these unwanted noise artifacts from the copper wiring. What does this have to do with wiring and magnetic fields? The reason is that any current flow creates an associated magnetic field. You cannot have one without the other. Therefore, this high-frequency unwanted signal creates a magnetic field that can interfere with surrounding equipment. It is the filter's function to remove this current so that its associated magnetic field will not interfere with other systems. This noise can originate either from the line or from the associated equipment that the filter is built into (load). From the equipment side, or load, the noise could be coming from computer clock frequencies, parasitic oscillations due to power switching, diode switching noise, harmonics of the line frequencies due to the high peak current charging the power supply storage capacitor, and many other sources. From the line, the noise could be due to flattening of the sine wave voltage caused by the high peak currents slightly ahead of 90 and 270 degrees due to the total of the power supplies fed from the line without circuitry to correct for power factor. This generates odd harmonics that feed the EMI filter. Other sources of noise from the line include equipment without any filtering and heavy surges of equipment being turned on and off. Lightning and EMPs (electromagnetic pulses) create other line problems for the filter.

To review, EMI is any unwanted signal from either the power line or the equipment, and this must be removed to prevent a magnetic field from interfering with closely associated equipment or to stop a malfunction of the equipment containing the filter. For example, it would be unacceptable for a patient's heart monitor to see degraded performance every time the local X-ray machine was used just because the same copper connected them. Here, the heart monitor filter would remove the pulse from the X-ray machine. Or, better yet, the X-ray machine input power filter would attenuate the noise to a level that makes it too small to be of concern.

2.2 Regular Filters versus EMI Filters

Most of the energy in the stop band (the frequency area to be attenuated) of the filter is reflected to its source. This fact is often overlooked in both standard-filter and EMI technology. The remaining energy is expended in the inductors through the DC resistance of the coil, the core losses (eddy currents and hysteresis), and the equivalent series resistance of the capacitors. All engineers have learned this in the past, but they often forget it somewhere along the way. Whereas this handicaps the regular filter designers, it is an aid to the EMI filter designer.

Standard filter designers have several excellent filter responses to choose from, namely: Butterworth, elliptic, Chebychev, and M-derived. They know the input and output impedances of the source and load (usually the same), the allowable passband ripple, the –3-dB or half-power frequency, and the stop frequency (the first frequency with the required amount of loss). The regular filter designer is able to use software filter-design tools or may even develop the filter design from mathematical derivation techniques using pole placement and ladder synthesis. In either approach, the filter performance should be very close to the results required. In some cases, the filter may have to be altered by adding stages to accommodate changes in topology, etc., to achieve the desired results. Needless to say, this type of filter typically demands equal source and load impedances and, therefore, is very deterministic, whereas the filter transfer function and the response required can be modeled with load and source normalized to unity.

In contrast, most EMI filter manufacturers design only the low-pass filters (all-pole networks) needed for the required EMI attenuation. Rarely do they build bandpass or other conventional filters. EMI filter design is not a precise process compared with that used by the conventional filter manufacturer; therefore, EMI filter component values are very flexible, allowing the use of standard values. These filters are adjusted only to meet the required insertion-loss specification, assuming that the rest of the specification is met.

The languages spoken by the two groups are also different. As mentioned in chapter 1, true filter houses often speak of poles, zeros, group delay, predistortion, attenuation, and terms such as the order of the filter. The EMI filter designer thinks in terms of attenuation, insertion loss, filter voltage drop, stability, and the number of filter sections required to meet the insertion loss. Although the power source may have harmonics, the actual power supplied to the device through the filter is restricted to the fundamental frequency. Such a harmonic content is especially true for power supplied locally by shipboard generators and remote stations where the generator is near or well past peak power. In this case, flat frequency response, low phase distortion, or low peak-to-peak ripple across the filter passband is not an issue here. These power line harmonics furnish no power to the load, so the EMI filter designer is not concerned with them. As a result, terms such as *group delay*, *ripple*, and *phase distortion* are never heard.

To summarize, the requirements of the conventional, or wave, filter house are entirely different from EMI requirements, as the technologies and levels of performance are completely different. The conventional, or wave, filter component values are more critical, and a need for precise values is paramount to filter performance and in meeting requirements.

2.3 Specifications: Real or Imagined

Specifications are another subject that often creates uncertainty and controversy. Some test specifications unnecessarily complicate the design and make it overspecified. For example, one company had been using a particular filter for years without any problems meeting the EMI qualifications. Then the test specification was changed from 220 A to the current-injection probe (CIP) method, and the filter never passed. These specifications are such that they conflict with reality. The 220-A specification calls for losses within the filter, with a source and load impedance of 50 ohms. The filter will, in reality, feed a power supply that is rarely close to 50 ohms and work into a source of rarely 50 ohms. In this example, the system was targeted for aircraft where the power feed would be very short. For this line impedance to approach 50 ohms, the line frequency would have to be in the MHz range. So what is the EMI filter designer to do: Match the 50-ohm specification or meet the real-world specification? What is meant here by the real-world specification is similar to the qualification test that may follow. This CE test measures the conducted noise that is generated within the system through the filter. The line-impedance stabilization network (LISN) is often used as the source impedance for these tests and is closer to the real-world requirements.

Most specifications that call for a LISN require 50-ohm output impedance. Unfortunately, the 50-ohm impedance is not reached until well above 100 kHz. If the only real issue is to match the 50-ohm impedance, the filter will be matched to the source impedance of 50 ohms. There are two concerns here. Firstly, what is the lowest frequency of interest and filter loss required? Is it below 100 kHz? The LISN output impedance drops rapidly from 50 ohms, and the filter is then mismatched. Secondly, what happens in the real world when neither the source nor the load impedance is close to 50 ohms? Figure 2.1 shows a typical LISN where the line impedance, whatever it is, is shunted by the input network at frequencies above the point where the 22.5-μF capacitor's impedance is equal to 1 ohm. This is at 7 kHz, so at frequencies above 10 kHz, the line impedance is reduced to the 1-ohm value, while the load side of the LISN looks like 50 ohms. The impedance of the inductor is low at 10 kHz (3.5 ohms). At 10 times these frequencies, the inductive reactance is only 35.2 ohms and reaches 50 ohms at 142 kHz. The entire network at 142 kHz generates a source impedance for the filter of 35 ohms at 44 degrees. It is obvious that the LISN will not look like 50 ohms until well above 150 kHz.

The proposed MIL-STD-461 LISN is shown in Figure 2.2. The real intent of the filter is to attenuate conducted emissions of differential- and common-mode origins from both the device and the line. The test specifications rarely prove that the filter will pass with any degree of satisfaction within the system specification or real-world specification. The filter can often pass the insertion loss in the test laboratory and fail when tested

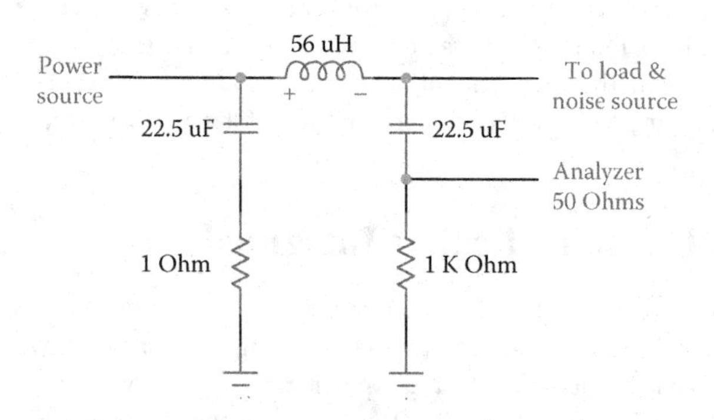

FIGURE 2.1 Typical LISN.

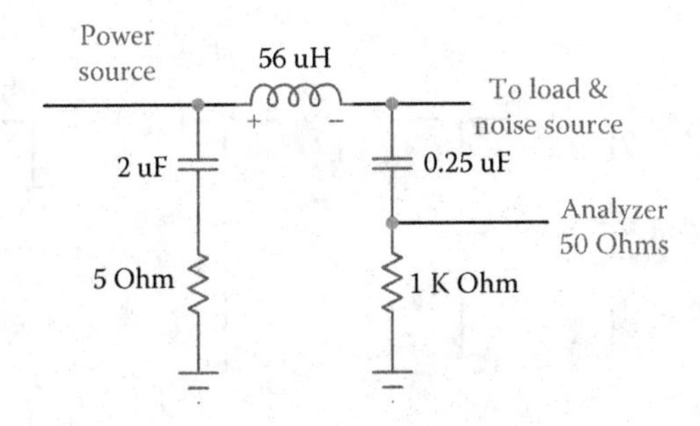

FIGURE 2.2 Proposed MIL-STD-461 LISN.

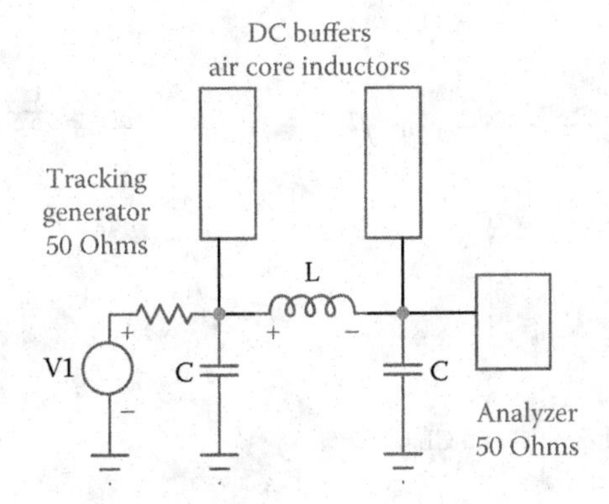

FIGURE 2.3 220-A test specification showing the buffers, if needed.

along with the system. A filter that has passed the full test as a target, or bogie, often gives disappointing end results. If the filter in question appears to pass the bogie, it is also possible that it will later be tagged as bad by the system tester.

The MIL-STD-461 specification is more realistic than the 220-A specification. Robert Hassett (vice president of engineering at RFI Corp., retired) has given several presentations for the Institute of Electrical and Electronics Engineers (IEEE) and other groups that show the advantage of moving away from the 220-A test method to the CIP method. Hassett has tested many filters by both test methods. The 220-A test setup is shown in Figure 2.3, and the current-injection probe (CIP) is shown in Figure 2.4. The Hassett curves show the difference using an L filter with the capacitor facing the line as compared with the inductor facing the line (Figure 2.5). The same is also true for the π filter due to its input capacitor facing the line. Either filter will look good under the 220-A tests.

These results are due to the 50-ohm source impedance. The CIP utilizes a 10-μF capacitor, and this will reduce the filter loss by 6 dB, especially at the lower frequency end. In the case of Figure 2.2, the signal generator—now normally from the spectrum analyzer tracking generator—has an output impedance of 50 ohms and feeds a coaxial switch (not shown), and the load impedance is the receiver's input impedance (also 50 ohms)

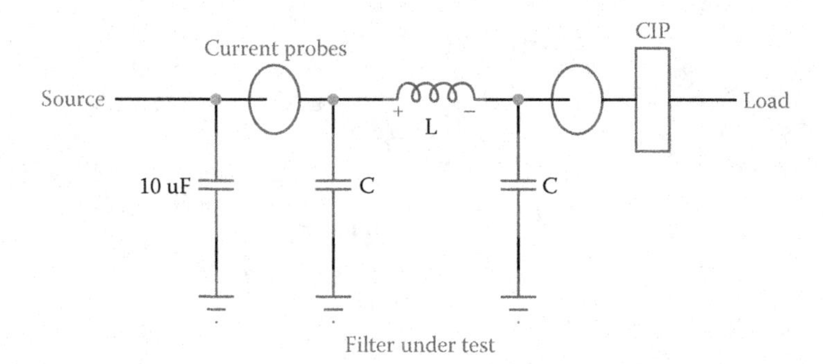

FIGURE 2.4 Current-injection probe method.

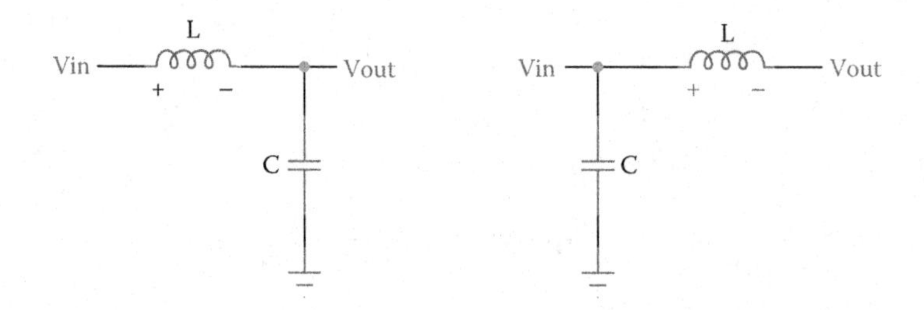

FIGURE 2.5 Two L filters, LC and CL.

fed through a second coaxial switch. A calibration path is provided between the two coaxial switches. This test method makes the π filter function as a true three-pole filter, giving 18 dB per octave or 60 dB per decade loss. The real-world initial losses at the low-frequency end would shunt out the input filter capacitor. This would give initially 12 dB per octave of loss until well over 100 kHz (depending on the line length). This test method masks this flaw of the π filter or the L filter with the capacitor facing the low impedance. Figure 2.4 shows the method used by Robert Hassett at RFI. In the MIL-STD-461 specification, the current probes feed the measuring equipment to compare the two currents. This method shunts the input π filter capacitor copying the real world up to the frequency where the 10-μF capacitor's SRF takes effect. This method makes the π filter loss 12 dB per octave or 40 dB per decade, instead of 18 dB per octave or 60 dB per decade.

Other test methods have been suggested. Mitchell Popick (vice president of engineering at Axel Corp., retired, and a member of SAE EMI group and the dB Bunch), recommends that the load side of the filter should face a diode bridge that is properly loaded. This is much better for those that will feed a power supply, which is about 95% of the time. This also shows how the filter handles diode noise, which is the leading noise competing with the switcher noise. These are primarily the odd-order harmonics of the power line frequency and spikes during turn on and turn off. It works well in three-phase systems as well, but it is important to make sure that the total inductive reactance

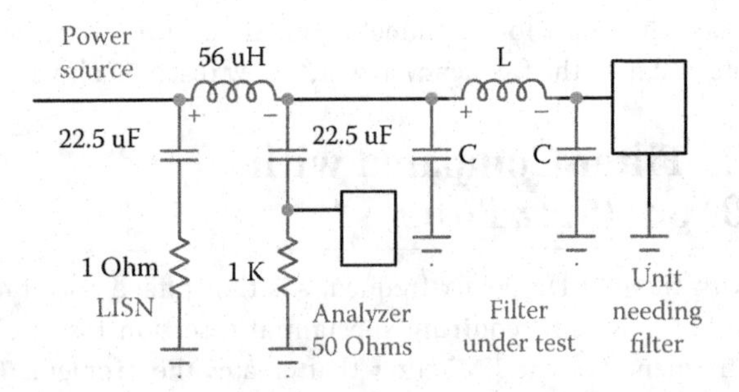

FIGURE 2.6 Analyzer method.

of the filter inductors is much lower than the primary inductance of the transformer (if any). This is especially true when multiphase transformers are to be used. These are often autotransformers, where the primary inductance is much lower than for the isolation transformer type. The inductance of the filter and the primary inductance of the transformer can form a voltage divider that reduces the voltage feeding the load. This is another reason to avoid the commercial filters, where one filter fits all. These may work fine in some applications and fail in others.

In Figure 2.6, the spectrum analyzer sees the diode, power switch, and parasitic noise that is allowed to pass through the filter under test. If the load is the real system rather than a load resistor, as in the drawing, the analyzer will see the full noise signature of the equipment. This is much more of a real-world test because the filter must handle all the noise sources at the same time plus the power for the unit, which is what happens in the real world. In other words, the filter could saturate under this condition, while the CIP and 220-A specification methods are looking at a single frequency from the tracking generator or CIP. Of course, both the 220-A and CIP tests pass the full power from the line, but so does the analyzer test method.

Reviewing, it is prudent not to use a capacitor input filter for the CIP test method because the loss is 0 dB per octave for this component, whereas it should be 6 dB per octave. This component costs money, demands filter room, and adds weight without performing until the frequency is very high.

2.4 Inductive Input for the 220-A Test Method

This is similar to the capacitor of the π filter in the CIP method, but not as severe. With the 50-ohms impedance in the 220-A test system, what is the inductor impedance going to add? At least the 50-ohm impedance is there, and the inductive reactance adds to it at 90 degrees. We are speaking here of either an L or a T filter. They are not responsive until the impedance of the inductor reaches 50 ohms. Regardless, this takes effect orders of magnitude ahead as compared with the capacitance to ground in the CIP method. Both L and T perform very well in the CIP method, but are somewhat limited in the 220-A method. If we were to calculate the frequency at which the inductor is 50 ohms, this would be the starting point where the inductance will start to function and explains why most filters are the π type required to pass the 220-A specification.

Summarizing, with regard to the inductors of the L or T filters in the 220-A test method, evaluate and note the frequency at which they reach 50 ohms.

2.5 400-Hz Filter Compared with the 50- or 60-Hz Filter

The problem with the 400-Hz power frequency is the voltage rise at 400 Hz. Again, we are speaking of a system requiring substantial insertion loss and load current. It is the cutoff frequency of the EMI filter that creates the problem. There is always a substantial voltage rise ahead of this cutoff frequency that pushes up the frequency spectrum at 400 Hz, and this creates a severe voltage rise at 400 Hz. At 60 or 50 Hz, the rise is so much smaller that other factors will compensate for it, thereby ensuring that the output voltage is the same as the input, with a potentially small voltage drop. This will be discussed later in the book, but the main answer is getting the cutoff frequency as high as possible for 400-Hz systems. This requires multistage filtering and impedance matching. As the number of stages grows, the cutoff frequency increases for the same amount of loss (80 dB at 100 kHz), thus dropping the gain at 400 Hz. However, as the current level decreases through the filter, this often enhances the voltage-rise problem.

There is also a technique referred to as RC shunt that decreases resonant rises due to circuit Q. For example, at 6 kHz, the resonant rise could be 10 to 15 dB. The resistor could be 10 ohms (covered later in the book), and the capacitive reactance at 6 kHz would be 10 ohms. Would this lower the resonant rise? Of course it would, but now you have 5 dB at 4 kHz. So what is the gain at 400 Hz? It is about the same as before, but there are cases where this fix made the condition worse. The resonant frequency dropped in dB but was also moved to a lower frequency, negating the fix. Another method is the Cauer, or elliptic, filter. In a multistage filter one (or two) of the inductors is tuned to, for example, 100 kHz. This adds many dB of loss, allowing all the filter component values to decrease until the filter loss is brought back to the needed 80 dB at 100 kHz. This pulls the resonant rise frequency farther away from the 400 Hz and reduces the voltage gain at 400 Hz. Of course, each change in the inductor values requires a change in the tuning capacitor so that it still resonates at 100 kHz.

Reviewing, 400-Hz filters with high loss requirements at low frequencies demand special handling to get the loss low enough and still have little voltage gain at 400 Hz.

3 Common Mode and Differential Mode: Definition, Cause, and Elimination

There is a wide-ranging set of opinions about the definition, cause, and elimination of both common-mode and differential-mode noise. This chapter attempts to cover most of these concepts.

3.1 Definition of Common and Differential Modes

Normal-mode noise is simply a voltage differential that appears briefly between the power line and its accompanying neutral or return line. The neutral line may not be power ground. As the name implies, these two lines represent the normal path of power through electric circuits, which gives any normal-mode transient a direct path into sensitive circuits and therefore the opportunity to degrade system performance. Normal mode is most often called differential mode.

Differential-mode noise is most often attributed to power supply or load switching. Other causes are transients, surges, or interrupts that occur on any line with respect to a ground reference. Major contributors of differential-mode noise are power supplies and motor controls that operate under PWM control where loads are switched at high frequency. The switching action creates differential-mode noise at the source due to high ripple currents in the DC link capacitors.

Common-mode noise is most typically seen as a transient voltage differential that appears between ground (not necessarily the neutral line) and both of the two normal-mode lines. Therefore, the noise is common to all lines with respect to a neutral or ground reference. Common-mode transients are most often the cause for concern, particularly so with sensitive analog and digital circuits, as this noise often leads to susceptibility issues where circuits can operate erratically, leading to adverse functional behavior or even failure. Common-mode noise impulses tend to be higher in frequency than the associated differential-mode noise signal. This is to be expected, as the majority of the common-mode signals originate from capacitive coupling of differential-mode signals. The higher the frequency, the greater is the coupling among the conductors, line,

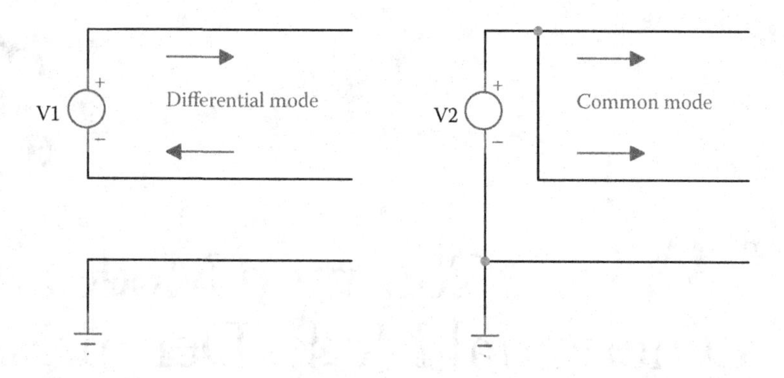

FIGURE 3.1 V1 (differential mode) and V2 (common mode).

neutral, and ground. Electronic circuits are typically much more sensitive to common-mode noise than differential-mode noise. To sum up, a differential-mode noise voltage is impressed between the lines, whereas the common-mode noise is seen across the lines—typically two—and ground (Figure 3.1).

3.2 Origin of Common-Mode Noise

The simple definition of common-mode noise is a signal that appears common to two or more lines relative to ground. This signal, or noise, is also conducted in the same direction. For the purpose of discussion, we consider the common-mode effects of a DC-DC converter, or a PWM switching supply.

A flyback DC-DC converter is comprised of a power switch on the input that chops the DC input voltage into an AC signal that is transferred across a transformer. The AC is then rectified back into DC for the output (Figure 3.2). The transformer is made from concentric windings that lie on top of each other for the primary and secondary winding structure; therefore, the transformer will have significant interwinding capacitance, including capacitance between input to output on the transformer.

Each time the power switch is in conduction, a large dv/dt is impressed by the input stage across the transformer input-output capacitance. This, in turn, causes a current to want to flow from input to output through the transformer capacitance. This current flows twice each switching cycle, and it must find a path back to the input "source." The current is commonly called the common-mode current, as it can flow through any or all of the inputs and outputs, either individually or at the same time (Figure 3.2).

An ideal transformer is a notional perfect circuit element that transfers electrical energy between primary and secondary windings by the action of perfect magnetic coupling. The ideal transformer will only transfer alternating, differential-mode current. Common-mode current will not be transferred because it results in a zero potential difference across the transformer windings and therefore does not generate any magnetic field in the transformer windings. Any real transformer will have a small, but non-zero capacitance linking primary to secondary windings as shown in Figure 3.2. The capacitance is a result of the physical spacing and the presence of a dielectric between

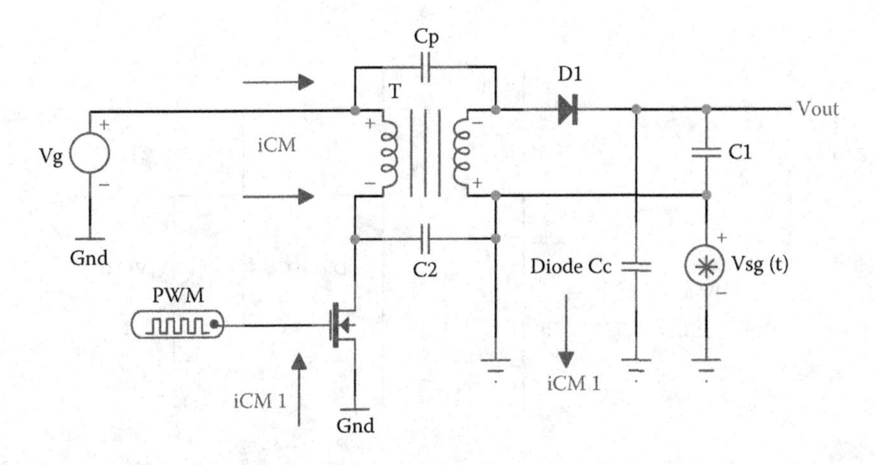

FIGURE 3.2 Isolated DC-DC converter (flyback).

the windings. The size of this interwinding capacitance may be reduced by increasing the separation between the windings, and by using a low-permittivity material to fill the space between the windings. For common-mode current, the parasitic capacitance provides a path across the transformer, the impedance of which is dependent on the magnitude of the capacitance and the signal frequency. In some cases, this common-mode current will have different magnitudes, depending on the parasitic inductance and capacitance, and may also contribute to the differential-mode noise. With no external current path from input to output (i.e., the converter is driving an isolated load that has very little capacitance back to the input), the common-mode current is largely contained within the converter and flows through the stray capacitance from the input to the output, therefore causing no further problems.

When an external current path is introduced to the converter, such as a PCB trace, as in the case of a nonisolated topology, or when an electrical connection exists between both input and output grounds, the current will tend to want to flow through this lower impedance connection (Figure 3.3). Typically, this is acceptable as long as the parasitic inductance of this connection is very low. The time-varying current in this conductor is di/dt, and is caused by the effects of power switching, or dv/dt. If there is appreciable inductance in the external path, the rapid change in induced current, or di/dt, will cause a voltage $V = L(di/dt)$ to be developed between the grounds. This will show up as voltage noise at either the input or output terminals. Therefore, grounds must be solid or low inductance, and loop areas must be minimized to counter these effects.

The transformer interwinding capacitance causes currents to flow between the isolated (primary and secondary) sides of the transformer, and can cause the secondary side ground voltage (V_{sg}) to switch at high frequency due to the high-frequency component.

In another scenario for common-mode effects, we may consider an indirect lightning strike and the effects placed upon equipment used within utility power distribution. The lightning strike will create a large magnetic field that couples into all power lines relative

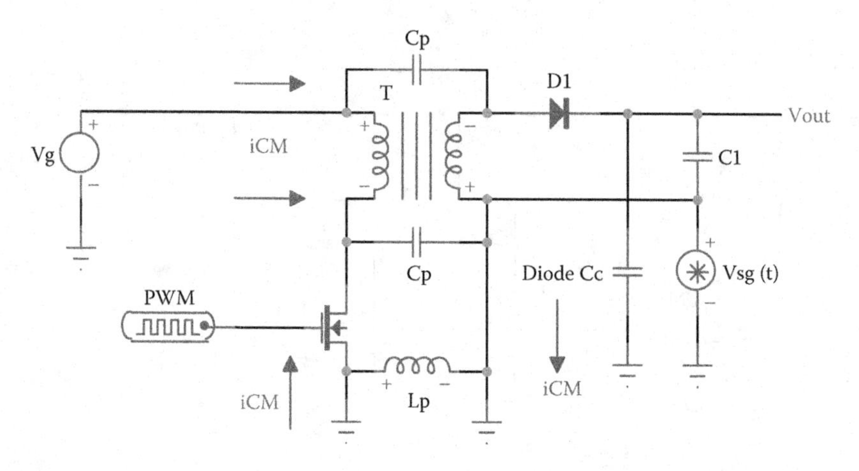

FIGURE 3.3 Nonisolated DC-DC converter (flyback).

to ground potential (Figure 3.1). This induced voltage is impressed between all conductors and ground and creates a potentially very high transient current flow in each of the conductors, and in the same direction. If the equipment does not have a method of suppressing this high voltage and current, it is likely that the equipment will suffer catastrophic failure. This strike could also include multiple bursts, or several high-voltage pulses, typically at approximately 50 kHz. The spacing between the lines may be very close as in the case of a power cable, or they may be farther apart. If the conductors are far enough apart, the magnetic field coupling will be somewhat different, thereby creating a slightly different induced voltage in the two or more power lines. In certain cases, some of these voltages may cancel out. During a strike, the induced voltage will be added algebraically to the AC power line voltage on all the lines. This section assumes that the lines do not fuse and that transformers are able to handle this pulse without failing. If the two impressed voltages are different in magnitude, the difference in these two line voltages feeding the transformer will be transformed to the secondary. This difference is now differential-mode noise, which is transformed (stepped down) to the secondary side. There will be extra transformer losses due to the high-frequency core losses as the induced noise, or voltage pulses, are at higher frequencies, thereby accentuating core losses. The skin effect in the transformer windings and on the lines also adds to the pulse losses.

The primary-to-secondary capacitance of the transformer (Figure 3.4) carries the common-mode current. However, the secondary has many paths to ground through the winding capacitance to the center taps at the transformer. The capacitance between primary and secondary is further reduced if the transformer has a Faraday shield or screen. The high frequency of the pulse is further reduced by the interwinding secondary capacitance of the transformer.

Most of the voltage transferred from the primary to the secondary through the capacitance will be shunted to various service grounds (Figure 3.5). It is primarily the difference voltage across the transformer that will carry the pulse to the utility service users.

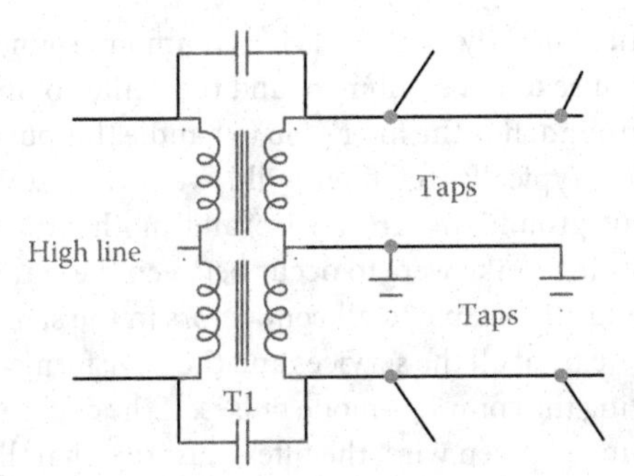

FIGURE 3.4 Line transformer with output taps with primary-to-secondary (P to S) capacity.

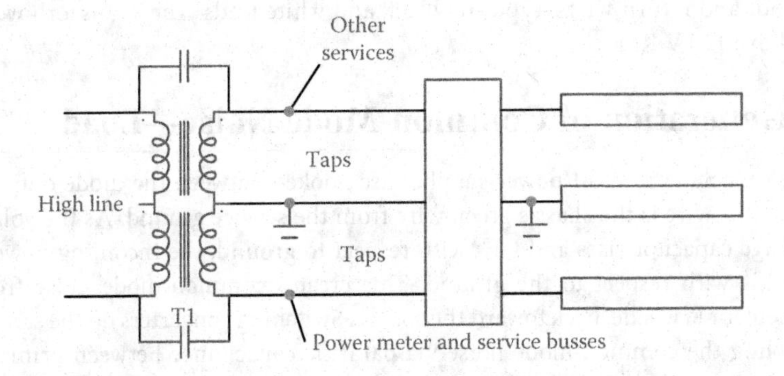

FIGURE 3.5 Typical service with power line, meter, and busses.

However, this voltage is differential mode and is carried by the two outer legs. Often, this voltage from the power company is called two-phase because it is 180 degrees out of phase. If the electrical equipment is connected across the two outside lines (220 V), the filter must handle a common-mode pulse. Both lines from the power panel are black (sometimes two different colors are used), and a safety green wire goes back to the service ground. Usually, the filter will have three Transzorbs (see chapter 12), with one from line to line and the other two from each line to equipment ground and carried by the green wire. These Transzorb devices will have two different ratings. The line-to-line Transzorbs will be rated at 250 or 275 V RMS, and the two line-to-ground Transzorbs will be rated at 150 V RMS. The Transzorbs are rated, or listed, by their RMS value. A Harris Transzorb V150LA20B has a rating of 150 V RMS and will fire at voltages around 212 to 240 V. The purpose of the arrester is to eliminate these high-voltage pulses. The line-to-line arrester eliminates differential mode, and the line-to-ground arresters eliminate common mode and help with the differential-mode noise.

If the equipment is powered off either side to the central ground, the pulse is then differential mode. Most equipment has three lines to it from the service: the hot wire, typically a black wire tied through the circuit breaker at the service; the neutral, typically white, tied directly to the service common ground; and the safety ground, typically green, also tied directly to the service common ground. The hot black wire carries the unwanted pulse to the equipment and the white common ground carries it back to the common ground. It is the filter's job to handle this pulse of noise that is now differential-mode noise. Typically, the filter will have a Transzorb from the hot to the filter case, or equipment, ground (the green wire) and another tied between the hot wire and return. If the lightning strike were to occur between the transformer and the user, the magnetic field would still couple to all conductors in the same way, but the central wire is grounded repeatedly at all the services and the transformer. In this case, the two outside lines are carrying the common-mode pulses. If the equipment is tied to the outside lines and the grounded green wire, the filter must then handle the common-mode noise artifacts. The three Transzorbs will be sized as before, with two rated above 120 V RMS and the one from line to line rated above 220 V RMS, typically 250 V RMS. If the equipment is tied from one line to ground, the noise energy is differential mode carried by the hot and return wires, typically black and white leads. The Transzorb would be rated above 120 V RMS.

3.3 Generation of Common-Mode Noise—Load

Storage capacitors in most power supplies are hooked between the diode outputs and ground. This wire is the chassis green wire from the service ground. As the voltage on the storage capacitor rises and falls with respect to ground, the incoming power lines follow this with respect to the ground. This creates common-mode noise from the equipment, or load side, back toward the source. Switching converters do the same thing by coupling the common-mode noise via parasitic capacitance between primary and secondary in the transformer. Power switches to ground also have relatively high parasitic capacitance, and the *dv*/*dt* on the switch will create *di*/*dt*; common-mode current will flow to switch ground. In some cases, an input transformer would eliminate common mode, and so would a power factor correction coil connected from line to ground. Isolating the input supply from ground by placing the storage capacitor across the diode bridge, and then following with an isolated switching converter, also works to remove the common-mode noise. Figure 3.6 shows the isolated supply with the storage capacitor

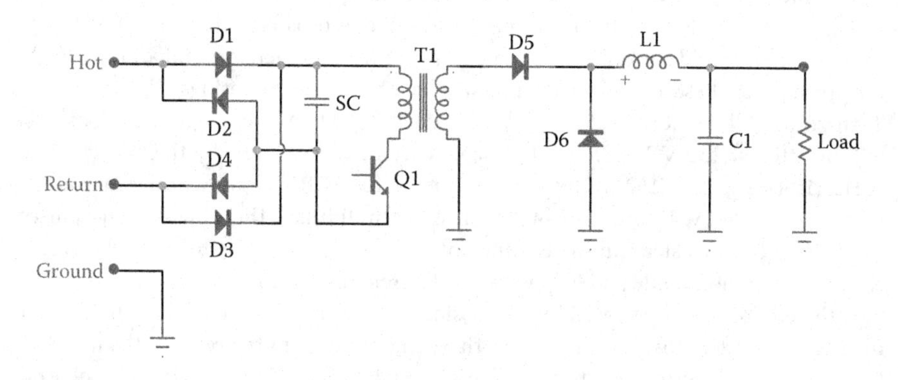

FIGURE 3.6 Simplified isolated input power supply.

(SC), the switch (Q1), and the load resistor (R1). As the transformer helps to eliminate the common-mode noise, the ground (green) wire has little current on it, referred to as leakage, or reactive, current. The EMI filter in front of this supply may be balanced with the differential capacitors connected from line to line, not to the green ground line, and leakage current will be minimal. Increased levels of common-mode loss is required if the system is powered from the 220-V side; however, larger capacitors to ground may be used because the system is balanced. See sections 3.4 and 3.5.

3.4 Elimination of Common-Mode Noise—Line and Load

Because common-mode noise is measured between both or all lines and ground, capacitors to ground are required to act as low-impedance shunts. In addition, common-mode chokes are used to present high impedance to common-mode current; these are discussed later. The reactive capacitor current to ground is also called leakage current. This ground current is the difference in current in the two ground capacitors from both sides of the lines. An isolation transformer eliminates the leakage current and greatly reduces the common-mode noise. Figure 3.7 shows two capacitors to ground for 220-V AC balanced lines. These capacitors could be feed-through type or leaded capacitors. If the voltage is equal and opposite (180 degrees out of phase) and the capacitors are equal, the ground current is zero at the line frequency. This works for capacitors with leads as well as for feed-through capacitors.

Earlier in this chapter, we said that there is no such thing as 100% differential-mode balance; therefore, there will be some current that flows to ground. An example would be 115 V from line to ground, 230 V line to line, 5% capacitor tolerance, 5 mA allowed to ground, and 60 Hz. Note that if the capacitance is a limit specified by a requirement and a capacitor measurement is taken from either line to the common ground, this method will not pass. If the current limit is specified, isolate the filter and load from

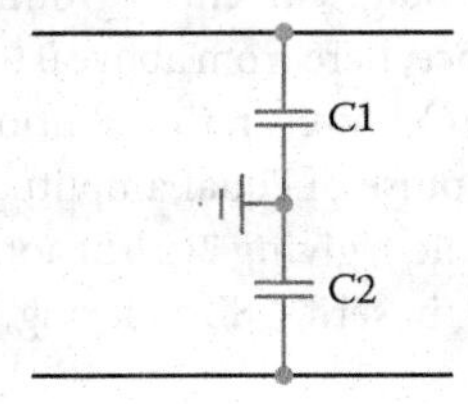

FIGURE 3.7 Balanced line-to-line capacitors for two-phase.

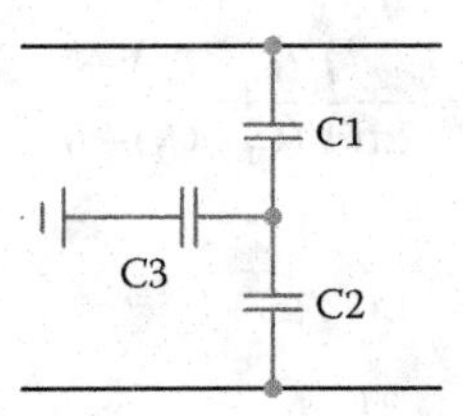

FIGURE 3.8 Balanced line-to-line capacitors with additional C3 capacitor.

ground and measure the green wire current, and this method will pass. The current through either capacitor to ground is

$$\begin{aligned} I_1 &= 2V\pi FC_1 \\ (I_1 - I_2) &= 0.005 = 2V\pi F(C_1 - C_2) \end{aligned} \tag{3.1}$$

The difference in the capacitors is double the percentage, here 0.1 C (one 5% high and one 5% low), and changing the capacitor value to μF and substituting the 0.1 C

$$\begin{aligned} \frac{0.005 \times 10^6}{2V\pi F} &= \frac{5000}{2V\pi F} = 0.1\,\mathrm{C} \\ C &= \frac{5000}{0.1 \times 2 \times 115 \times \pi 60} = 1.15\,\mu\mathrm{F} \end{aligned} \tag{3.2}$$

Therefore, two feed-through or leaded capacitors of 1 μF will work well if the capacitor tolerance is 5% or less. Remember, the values used in this configuration work only in the line-to-line system, and a much smaller single value of capacitance to ground would be required in the 120 V–to-ground arrangement to meet this low current value.

Furthermore, this approach should never be used for medical equipment where a patient may be connected to the equipment. If one of the wires opens, the full line-to-ground voltage (120 V) of the one remaining line is impressed across the one capacitor, and the current to ground through it is 45 mA, well above the patient limit.

Another arrangement for the two-phase balanced system, which also eliminates using the feed-through capacitors, is to replace the ground at the common point with a capacitor to ground (see Figure 3.8). This is used for common-mode attenuation. Here V is the line-to-ground voltage, C_1 and C_2 are the line-to-common capacitors (assuming C_1 is the larger of the two), C_3 is the junction-to-ground capacitor, e is the junction voltage, and I_1 is the maximum leakage current to ground requirement through C_3. The equations follow. If t is the tolerance, here from above 0.05 tolerance, then the maximum difference in the capacitance is $2tC$; however, the addition of these two is assumed to be 2C. Assuming a common-mode pulse of equal amplitude on both lines to ground, the two lines to junction add (in parallel), giving 2C. The total capacitance to ground would be the two in parallel and then C_3 in series. Simplifying,

$$\begin{aligned} e &= \frac{V(C_1 - C_2)}{(C_1 + C_2 + C_3)} = \frac{2VtC}{2C + C_3} \\ C_3 &= \frac{I_L(C_1 - C_2)}{2\pi FV(C_1 + C_2) - I_L} \\ C_3 &= \frac{2tI_L C}{4\pi FVtC - I_L} \\ C_T &= \frac{I_L}{2\pi FVt} \end{aligned} \tag{3.3}$$

The most practical solution along with the best overall performance is to have the three capacitors equal in value.

$$C = \frac{3I_L}{4\pi FVt} \quad (3.4)$$

Make C_3 equal to C in the preceding equation. Round down the three capacitors to a convenient standard value. It is important to ensure that the tolerance limit of the two line-to-junction capacitors is correct. The junction to ground is not as critical. Another method, but a costly one, is to sort, or grade, the capacitors into smaller difference percentages. Use the matched ones for the two line-to-junction capacitors and the oddballs for the junction to ground. A difficulty arises if someone uses this 220 line-to-line filter for a 120-to-neutral filter. Now the junction-to-ground capacitor and the junction-to-neutral, or the return capacitor, are in parallel, and the voltage from junction to ground is V divided by 3. In this situation, the current through C_3 to ground is

$$I_G = \frac{3VI_L 2\pi F}{3(4\pi FVt)} = \frac{I_L}{2t} \quad (3.5)$$

where I_L is the original design leakage current and I_g is the new ground current. It is assumed that the actual calculated values of the capacitors are used; otherwise, I_l is the resultant leakage current, which is lower than the specified value. In the preceding case, with I_L equal to 0.005 and t equal to 0.05, I_G is equal to 50 mA.

This is well out of specification. Hopefully, the leakage current is not specified for this requirement or is a larger value for this application. A 220-V AC balanced filter designed as shown here should be marked on all documentation, and the system should not be used for 120-V AC line-to-ground applications. If the system is being built in-house where the filter, power supply, and the rest of the system are under engineering control, build the filter in as part of the supply and design a current transformer for the ground lead of the capacitor. Use this to shut the system down if a voltage imbalance occurs or if there is excessive current to ground for any reason. Put this network reasonably close to the line input with a current transformer on either side of the capacitor to ground. Design this to operate a relay that opens the system after the filter where the relay is energized only when excessive ground current is detected.

Summarizing, the balanced line for 220-V AC for two phases with leakage current specifications can be met with the three-capacitor arrangement. This technique was developed for common-mode noise. The normal voltages are 180 degrees out of phase, whereas common mode is in phase. So the capacitors buck for normal mode and add for common mode. This adds to the loss of the common-mode inductor. It is not advisable to use this in any medical equipment that would make contact with a patient. One way to get around this is to use a current probe that would shut down the equipment while also opening any connection to the patient. If this cannot be achieved, mark all documentation to state that this system is never to be used on 120-V equipment working between line and ground or for any medical application. This application is similar to the virtual ground for three-phase systems.

3.5 Generation of Differential-Mode Noise?

The common-mode pulse was discussed in section 3.2. A transformer was placed between the lightning strike and the filter. For the most part, this creates differential-mode "energy" at the filter. Thus, it is recognized that common-mode noise can create differential-mode noise. We are not familiar with the reverse condition. The line side can create differential-mode noise when inductive equipment turns on and turns off. This is a voltage pulse between the lines. Likewise, transformers between the output of the filter and equipment produce differential-mode noise for the filter to handle. For the line side, MOVs, or Transzorb devices, help to clamp the higher voltage pulses from the line and equipment, and the differential filter section must handle the rest.

3.6 Three-Phase Virtual Ground

This technique can only be used for the three-phase virtual applications all within the same enclosure. Types requiring individual insert filters with all the capacitance to ground cannot use this technique. This is very similar to the preceding two-phase application. A capacitor is tied from each phase to a common point, making a virtual ground. If the voltage of each phase voltage is the same and the three capacitors are the same, then the junction voltage is zero, or a virtual ground. If a fourth capacitor is tied between the junction and ground, the current through this capacitor is zero. If the unit is tested for ground current—not capacitance to ground—by isolating both the equipment and the filter from ground, then measuring the current on the ground wire should indicate that the current is well below the specification limit. Again, the capacitor values are equal for best overall results. Solve equations similar to the preceding equations. The voltage at the junction is equal to the line voltage if one phase fails, and the phase angle will be between the two remaining phases—120 degrees from each. Therefore, this presents risk for the use of medical equipment involving application to a patient. However, three-phase high-power equipment would probably never be used in an application that involves a patient. Again, use a current probe to monitor the current and open a relay to remove the power from the equipment or the patient, or both. The latter statement assumes the design engineer has control over adding a ground fault device and is not just designing the EMI filter. This situation is similar to that discussed in section 3.4.

The best part about all this is that the three-phase voltages give nearly zero current to ground, but for the common-mode voltages, the three capacitors are in parallel, with the fourth in series. This gives very low impedance to ground to eliminate the common-mode noise. The technique discussed here, including a properly designed common-mode choke, will almost certainly eliminate any common-mode problem. It is important to remember: If there is a neutral wire, there should be equal windings for the three phases plus an additional winding for the neutral wound on the ferrite common-mode inductor. Any imbalance in the phase currents is carried by the neutral, so the total magnetic flux will be zero in the common-mode core. The theory is that the neutral current is low in a balanced system. Any multiple of the third harmonic that exists (3, 6, 9, etc.) in the system adds back in phase and adds to the neutral. So, the total third-order currents and the unbalanced current add to the neutral. Make sure the wire size can handle at least the peak phase currents. It is not uncommon to see neutral filters overheating well above the temperature of the other phase filters; this is due to imbalance in current and the third-order harmonics. Measure the current on the four wires to see if the neutral current is below the other three. The leakage current through C1, C2, and C3 is as follows, where E1 and C1 are, respectively, the voltage and the capacitor to the virtual ground for Phase A. E2 and C2 are for Phase B, and so on.

$$I_L = \omega\left[\mathrm{E1C1}\sin(\omega t) + \mathrm{E2C2}\sin\left(\omega t + \frac{2\pi}{3}\right) + \mathrm{E3C3}\sin\left(\omega t + \frac{4\pi}{3}\right)\right] \tag{3.6}$$

To review, what removes common-mode noise? The answer is a common-mode choke, capacitors to ground (feed-throughs or Y caps), transformers, and arresters.

4

EMI Filter Source Impedance of Various Power Lines

One of the leading questions asked by people who have EMI issues or who are joining the EMI field is related to source impedance and how this affects the frequency magnitude response of the EMI filter. Engineers who have experience with EMI filter design will want to have a good approximation of an EMI filter's transfer function *H*(*s*) that is being used in their application. This is also true if the filter is being developed by a third party. The reason behind this is to ensure that the filter is optimized for stability. Even if a filter is developed by a third party, it is possible to determine an accurate equivalent two-port transfer function that should yield half-power bandwidth, frequency magnitude attenuation, and the filter Q. In some cases, it may be prudent to measure the line impedance along with the load impedance and to know the transfer function of the third-party EMI filter so that an accurate assessment of insertion loss can be made. The transfer function can be derived; however, most EMI filter manufacturers rarely base their design approach on formal passive network analysis. If the filter manufacturer is able to provide this information, it is possible to add another section between the filter and the load. The purpose would be to notch out resonant rises and other instabilities of the entire system.

This procedure would work very well if the engineers were certain that the line length, conductor spacing, or diameter of the conductor would not change much from installation to installation. These parameters require precious time to determine and could require expensive rental equipment for most companies. This would work well if their system—including the EMI filter device—were to be installed on a particular type of equipment where cable dimensions are approximately the same. The technique would fail to work if the unit were to be installed on different equipment or in applications that have very unique wiring interfaces. Also, they would need to consider all possible load conditions, including the peak load. If the device would go into some standby mode, the instabilities could shift and the filter along with the rest of the equipment could oscillate. The harmonic content of the power line frequency varies from line to line. In the past, most commercial lines had little harmonic content because of the very low line impedance at these frequencies.

This was discussed with a large power company in Southern California in 1985. They stated that 85% of the power was used for "power and lights," that is, motors and lamps and, at that time, incandescent lighting. At that time, and with the proliferation of computers, computer printers, scanners, copiers, fax machines, televisions, sound systems, and similar equipment that were tied to these lines—and very few with power factor correction circuits—the voltage waveform was less sinusoidal. This was due to the voltage drop caused by the high current spikes that these machines demand a little ahead of 90 and 270 degrees. The power factor correction capacitors placed across power lines (static Var correction) by the power company did help to reduce the power factor angle with a corresponding reduction in Var loss; however, the voltage was still somewhat distorted with harmonic content. The harmonic content of generators at remote sites and shipboard installations is much greater because of the higher resistance of the generator and lines. The voltage supplied to the end users in these applications is less sinusoidal. A power consultant in Southern California found 100-A spikes above the nominal sinusoidal current in small office buildings that had power problems. Again, these spikes were a little ahead of 90 and 270 degrees, typically around 85 and 265 degrees.

4.1 Skin Effect

As the frequency increases on the line, the depth of conduction is reduced. The wire cross-sectional area decreases because the radius of conduction decreases. The higher the AC resistance is, the greater the dissipation of this unwanted energy. Skin effect can take its toll on the higher frequency energy on the power lines. This helps dissipate the electromagnetic pulse (EMP) and other higher frequency noise traveling on the power line in either direction. These power lines were constructed to handle power at very low line frequencies and not for the higher frequencies creating the loss. Although the characteristic impedance of the line may be 50 ohms, the loss of the line per unit length increases with frequency. For copper, the equation of the skin-effect depth in centimeters is

$$D = \text{depth(cm)} = \frac{6.61}{\sqrt{F}} \tag{4.1}$$

The cross section of the conducting area (CA) of the wire for frequencies above the skin depth is

$$\begin{aligned} \text{CA} &= |R^2 - (R-D)^2|\pi \\ &= D(2R - D)\pi \end{aligned} \tag{4.2}$$

where R is the radius of the wire in centimeters, and D is the skin depth from above, also in centimeters. As the frequency increases, D decreases such that the D term is much smaller than $2R$. Equation (4.2) at these upper frequencies is reduced to

$$CA = 2\pi RD \tag{4.3}$$

The original cross-sectional area, πR^2, compared to the value of CA multiplied by the original DC resistance, will give an approximate value of the AC resistance at these upper frequencies due to the skin effect.

$$\begin{aligned} R_{ac} &= \frac{\pi R^2 R_{dc}}{2\pi RD} \\ &= \frac{RR_{dc}}{2D} \end{aligned} \tag{4.4}$$

The lowest frequency would be well above the value of

$$F = \frac{6.61^2}{R^2} \tag{4.5}$$

Below this frequency, the skin effect is the full radius of the wire. The frequencies we are discussing here should be several times this lower frequency. Replacing D with its definition in equation (4.1)

$$R_{ac} = \frac{RR_{dc}\sqrt{F}}{13.22} \tag{4.6}$$

If R_{dc} is the resistance in ohms for a small distance along the line, then R_{ac} will be the approximate AC resistance for this short section. R_{ac} along with L, C, and G, the conductance across the line, will form a short segment of this line, also known as a per unit length. This has a characteristic impedance of

$$Z_0 = \sqrt{\frac{R_{ac} + j\omega L}{G + j\omega C}} \tag{4.7}$$

Substituting equation (4.6) into (4.7)

$$Z_0 = \sqrt{\frac{0.07564RR_{dc}\sqrt{F} + j\omega L}{G + j\omega C}} \tag{4.8}$$

Simplifying yields

$$Z_0 = \sqrt{\frac{0.07564RR_{dc}\sqrt{F} + 2\pi FL}{G + 2\pi FC}} \tag{4.9}$$

We may delete G because it is much smaller than $2\pi FC$ at the frequencies discussed in this section. The square root of F is removed in the first term in the numerator

$$Z_0 = \sqrt{\frac{0.07564RR_{dc} + 2\pi L\sqrt{F}}{2\pi C\sqrt{F}}} \tag{4.10}$$

To find the limit with respect to frequency (F), differentiate the numerator and the denominator separately

$$\sqrt{\frac{0 + \dfrac{2\pi L}{2\sqrt{F}}}{\dfrac{2\pi C}{2\sqrt{F}}}} = \sqrt{\frac{2\pi L}{2\pi C}} = \sqrt{\frac{L}{C}} \tag{4.11}$$

To summarize, the first term of the square root numerator vanishes and the term $2\sqrt{F}$ in the last term of the numerator and denominator cancel.

$$Z_0 = \sqrt{\frac{L}{C}} \tag{4.12}$$

Therefore, the skin-effect term has little or no effect on the characteristic impedance. The values of L and C are the dominant terms. This is the fundamental equation of the characteristic impedance of coaxial cables. This all shows that the normal characteristic impedance equation still dominates at the higher frequencies and that this characteristic impedance does not change with skin effect, although the loss per unit length does.

This behavior is similar to that of coaxial cable except that coax is designed to handle higher frequencies. There are many different coaxial cables with the same characteristic impedance. These lines have different inside and outside diameters, some with very small diameters and others with larger diameters. The dB loss per 100 feet varies from coax to coax and also varies with frequency. Think of it as a pad. The impedance of the pad may be 50 ohms, but the loss of the pad varies from pad value to pad value. The main difference here is that this power line impedance also varies with frequency. It is this line impedance that dissipates the unwanted energies, not the characteristic impedance of the transmission line.

Skin effect also applies to the wire used for the inductors, the transformers (if used), and the rest of the EMI filter wiring. The purpose of the filter is to rid the system of unwanted signals or noise. What better way is there than to dissipate it? Use the skin effect to do this. For the filter inductor, it is important never to use Litz wire or strands for the turns. Allow the wire to help dissipate these upper frequencies. The same is true for the rest of the hookup wire. Is there an exception? Yes. Some larger current filters have capacitors in parallel to ground. Here, Litz or braided wire should be used so that the capacitors can shunt these signals to ground. Otherwise, the capacitors will be limited in performance and have a lower self-resonant frequency (SRF) due to the lead inductance.

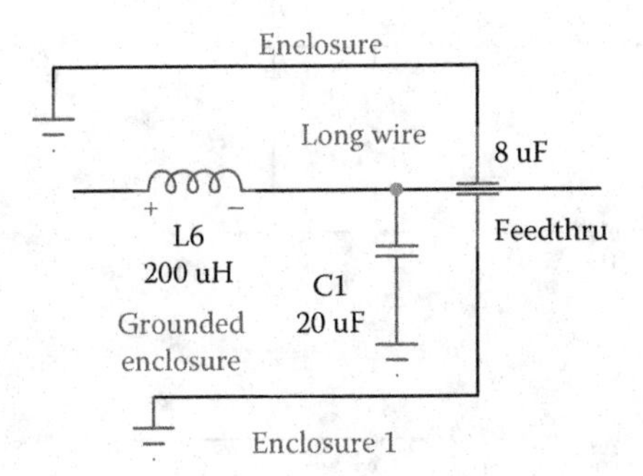

FIGURE 4.1 Long inductor lead example.

It makes no difference how long the inductor leads are, and often the lead is a single strand, as covered previously (Figure 4.1). The lead's properties ultimately add to the loss of the inductor with lead inductance and loss of skin effect. The high-current EMI filters (100 A and higher) require capacitors of 10, 20, and 30 μF and are soldered directly to the enclosure. These have higher working voltages, typically 100 to 440 V. The other capacitors are high-quality feed-throughs with values of 4 to 12 μF. These larger value capacitors are oval shapes that provide low-frequency loss in the 20-kHz range. The inductors are typically C-cores and are often multistrands of wire to carry the current. The point here is that the inductor leads can be self-leads and can be wired directly to the capacitor terminal while the soldered oval can is the other terminal. The wires are kept away from the enclosure, so there is little capacitance to ground this way, and the minor value of inductance along these leads just adds a small amount of inductance to the inductor, and does not reduce the SRF of the capacitors. The wires from the inductors are self-leads, but the wires to the oval capacitors should be better-quality braids. If the inductor self-lead is attached to the capacitor, this is fine; however, the capacitors to ground need the lowest impedance to work. This is why these capacitors are soldered directly to the grounded enclosure.

Leads with inductance and skin effect between capacitors lower the SRF of the capacitor, thereby degrading the performance. The inductor's longer lead is tied to a feed-through that is part of the output terminal. However, there is a second capacitor tied in parallel to the feed-through. This last lead should be a braided lead that offers a very low skin effect and low inductance. Even with a quality lead, this lead should be as short as possible. There is no perfect lead here, and the cure is to make these leads as short as possible. In practical terms, it is important to place this capacitor as close as possible to the other components in order to keep the tie leads short.

4.2 Applying Transmission Line Concepts and Impedances

Transmission line concepts in Figure 4.2 may sound like a strange subject to introduce when discussing EMI filters, but the filter designer needs to have a basic understanding of this subject for several reasons. The first is to understand why high impedance is

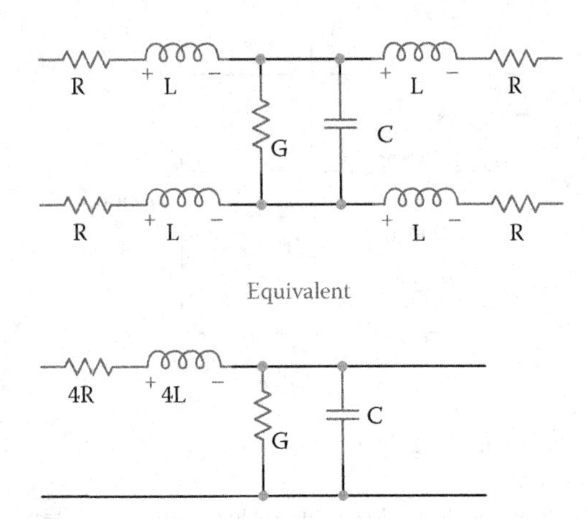

FIGURE 4.2 Transmission line and differential-mode equivalent for calculations.

needed at the filter input end for EMP applications. The second is to understand that the power line energy loss is due not only to the characteristic impedance of the line, but to the resistance elements along the line. These losses are due to elements such as skin effect, the DC resistance of the lines, and the conductance across the line. The third reason, especially in multiple-element filters, is that the filter has characteristics similar to those of transmission lines. The shorter power lines start taking on their varying characteristic impedance at much higher frequencies. The characteristic impedance of these power lines varies significantly; characteristic impedance varies with frequency and is not as constant as that of coax, twin lead, and twisted pair. The characteristic impedance of the open wire–type typically varies between 50 and 180 ohms due to the spacing between the conductors and the diameter of the wire.

The paired type, or twisted wires enclosed in conduit, are typically 50 to 90 ohms for the same reason, as previously stated, with the added capacitance between the wires and the conduit. The conduit adds little shielding because of the thinness of the material. In about 95% of the cases, several different power line sections will be in tandem. These different power line impedances "vee" back and forth as the different power sections approach resonant lengths. The velocity of propagation is very low due to the line being constructed to carry only the power line frequency, not radio frequency. The electrical lengths of these lines appear to be about eight times their actual length. If the power line is struck with a pulse, the power line will dissipate some of this energy in the resistive elements mentioned previously. The pulse will travel toward the filter end at the lower velocity of propagation of these cables. The fundamental frequency is around 50 kHz for lightning and soon assumes the characteristic impedance of the power line cable rather than the free-space impedance of 377 ohms. If the filter input impedance is high or looks like an open-impedance to this pulse, the voltage soars and could double, thereby aiding the arrester to function quickly. If the filter impedance is low, or is seen as a short to this pulse, the voltage drops to zero and the line current doubles. Some engineers report that some graphs from the IEEE literature have listed specific line impedance spikes of almost 500 ohms at one frequency in the megahertz (MHz) range.

The major contribution of these losses is due in part to the skin effect and the DC resistance, especially at the lower frequencies for the longer lines. Short lines appear to form link coupling. The point here is that at frequencies below 10 kHz, both the longer and shorter lines look resistive and close to the DC resistance value. This value is near zero ohms for the better commercial power companies, but for most remote power systems, the resistance can be much higher. As the frequency increases, these lines have output impedance that reaches 4 ohms at 10 kHz. The impedance then ripples its way to 50 ohms near 100 kHz for the longer lines and 250 kHz for the shorter lines. It should be obvious that the losses required for the Military Standard 461 (MIL-STD-461) specifications are such that the filter must be designed to have the proper losses with the line impedance in the very-low- or the zero-ohms region. This should make it clear why the pi filter has trouble meeting the loss for the real-world and MIL-STD-461 requirements. The same is true using a 10-μF capacitor across the power line in the common current-injection probe (CIP) testing. The losses that the filter must meet are at frequencies in the region 10 to 14 kHz in the MIL-STD-461 specification and may be as much as 100 dB for power line filters. This is the type of filter used in secure-room applications. The skin effect applies within the filter. Regardless of filter type, often many feet of wire is used, so the skin effect will dissipate the higher frequencies within the filter body.

The wires used for inductors are selected to be sufficient to carry the power line current at the power line frequency. The voltage drop and inductor-temperature rise will dictate the wire gauge used. Stranded wire is used only to ease winding techniques rather than for high-frequency loss considerations. In fact, these high-frequency losses are beneficial to high-frequency filter loss requirements. In addition to the skin effect losses, core losses at the higher frequencies are beneficial in increasing the filter losses. These are dissipated and not reflected to the source.

4.3 Applying Transmission Line Impedances to Differential and Common Modes

The transmission line losses for the differential mode, as in Figure 4.2, are all the resistive elements of the line, including the reciprocal of G, the conductance across the line. These losses include the skin effect and some very minor losses due to the equivalent series resistance (ESR) of the capacitor. The DC resistance (DCR) of the inductor adds no loss because the DCR is already included in the wire—it is the wire. The impedances all along the line in each section will dissipate any differential mode pulse, motor, or inductive load switch, propagating down the line toward the line filter.

The difference in the requirement for common-mode (shown in Figure 4.3) and differential-mode attenuation is that G and the ESR of the capacitor have no effect in the common mode. A pulse traveling down the line, lightning or EMP, toward the filter is partially consumed by the resistance of the line and the line's skin effect. Again, the DCR of the inductor(s) is included with the wire losses (this is the wire). The characteristic impedance also changes because the two lines are in parallel, so the capacitor across the line is out and so is G. The two inductors are in parallel, about 1.5 μH per meter

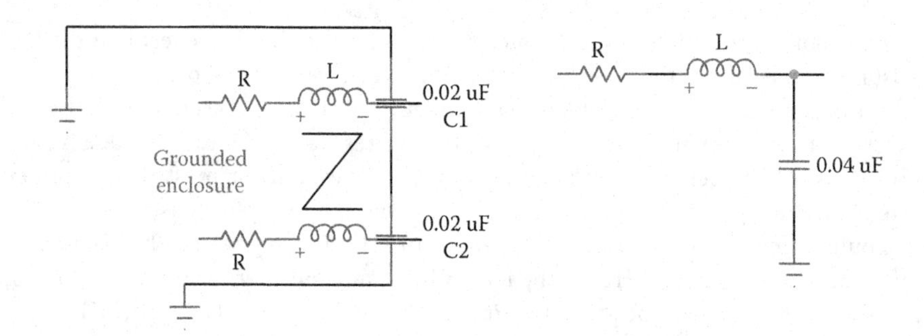

FIGURE 4.3 Common mode converted to differential mode for calculations.

on each line, or 0.75 μH, and the capacitance to ground determines the characteristic impedance. This makes the characteristic impedance quite high, which explains why the EMI common-mode filter impedance can be very high.

4.4 Differences among Power Line Measurements

Many groups have measured the line impedance, and there has been little agreement when comparing their results. The reason is that the lines being compared vary so much. The velocity of propagation is quite slow in all these lines, making the lines appear much longer than they really are. In some lines, the ratio is eight times their actual length. Devices such as generators or transformers terminate the line at the higher frequencies. This is due to the capacitance across these devices, which shunts the line at these higher frequencies. These lines were designed to carry DC, 50-, 60-, or 400-Hz power and not these higher frequencies.

4.5 Simple Methods of Measuring AC and DC Power Lines

Power line impedance can be easily measured, but this should be done with caution if the designer wishes to do so. There should not be any need to do this, especially when using the techniques used in this book. This measurement can be accomplished with inexpensive test equipment on both AC and DC lines, but it would be much faster and more accurate using a network analyzer. Measurement of either AC or DC line impedance should be resistive loaded, and several runs will be needed. The first readings are with maximum loading, followed by medium loading and then low loading.

To measure the AC line, start well above the line frequency by at least a factor of 10. This is done to keep as much of the AC line harmonic as possible to get more accurate readings. The frequencies required will be over a wider frequency range well into the kilohertz area. A frequency-selective level meter (FSLM) could be used along with a signal generator and a blocking capacitor. The capacitor should be high impedance to the line frequency and low to the frequencies to be measured. If the line frequency is 400 Hz, the lowest intended frequency to be measured should be well above 4,000 Hz. If the impedance of the capacitor is 1,000 ohms at 400 Hz, the impedance would be 100 ohms at the lowest frequency to be measured. The FSLM should be used in its narrowest input filter selection, if available. The generator frequency should still not be tuned to any multiple, especially the odd harmonics, of the line frequency until 20 times the line frequency is reached. The loss across the capacitor cancels except at the lowest frequencies; however, the 400-Hz loss could be still higher.

The only component remaining is a resistor of some value, say 400 ohms, and this should be noninductive to well above the highest frequency reading. The resistor, capacitor, and signal generator—in that order—are tied across the line with the resistor at the hot end and the signal generator to the neutral. The signal generator is tuned to the desired frequency; the FSLM has its low side tied to the neutral; and two points of measurement are taken. First, the voltage reading is taken between the capacitor and resistor where the FSLM is tuned to peak at the generator's frequency. The second reading is taken at the high line without readjusting the FSLM frequency. If a high-pass filter is added, a voltmeter could be used instead of a frequency-selective level meter. The readings will not be as accurate, but will be close enough.

Make sure that the FSLM front end can withstand this AC line voltage without blowing out. Many of the frequency-selective voltmeters or level meters have precision input pads that may quickly generate "blue and expensive smoke" if the hot line is touched. A high-pass filter with a series capacitor input, not an inductive shunt input, could be built in a probe to be installed at the test lead to protect the FSLM's input, as shown in Figure 4.4. The cutoff frequency of this filter must be at least half the lowest frequency to be measured.

This filter does not have to be special and does not require a flat Butterworth response filter, and it can tolerate ripple in the passband area of the high-pass filter. The filter errors are cancelled, as shown in equation (4.14). The ratio of the two readings will be the same because the errors cancel in the equation at the same frequency. As an example, at 400 Hz, the lowest reading should be 4,000 Hz. The reason for the multiple of 10 is to avoid the potential high-level harmonic content on some power lines. The high-pass

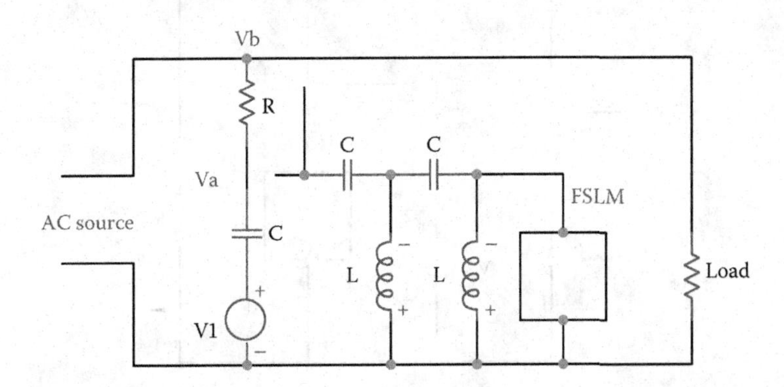

FIGURE 4.4 Measuring the AC source impedance.

filter should have a cutoff frequency of 2,000 Hz. Then the equations for the values of the inductors and capacitors in the high-pass filter are

$$L = \frac{Z_{\text{fslm}}}{2\pi F_0} = \frac{50}{2\pi \times 2000} = 0.004$$
$$C = \frac{1}{2\pi F_0 Z_{\text{fslm}}} = \frac{1}{2\pi \times 2000 \times 50} = 1.592 \times 10^{-6} \tag{4.13}$$

This assumes that the input impedance of the frequency-selective level meter is 50 ohms. Record both of the readings and move on to the next frequency to be measured. The formula is

$$Z_0 @ F_0 = \frac{V_b R_1}{V_a} \tag{4.14}$$

The voltage level V_a is recorded at F_0 between the capacitor and resistor, where V_b is the high line voltage reading. In addition, R_1 is the value of the series resistor between the high side and the capacitor. *F* is the high-pass filter installed ahead of the FSLM to protect the input pads. The impedances at these various frequencies can be plotted for the various loads and compared in an Excel spreadsheet. It is often better to use a battery-operated FSLM and signal generator to avoid ground loops.

The DC measurements are almost the same, except that lower frequencies can be read and the highest frequency needed will be in the lower kilohertz range (compare Figure 4.4 and 4.5). Obviously, the high-pass filter is not required. These resistive readings can again be plotted in a spreadsheet. The readings will level out (resistive) at a low frequency in the kilohertz range. Here, you could replace the FSLM with an AC

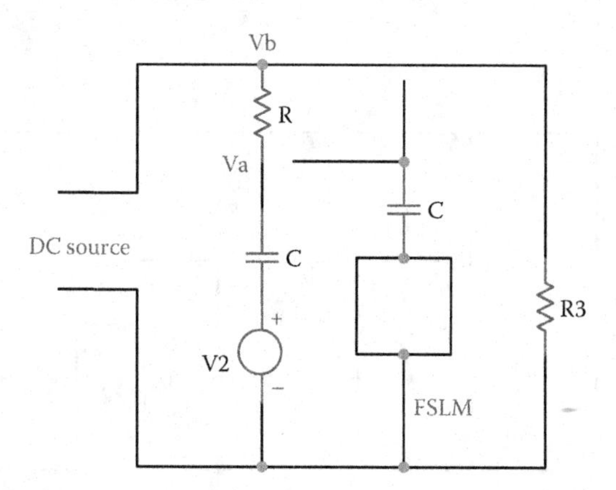

FIGURE 4.5 Measuring the DC source impedance.

voltmeter. This setup does not require as much caution as in the AC readings. Make sure, again, that your FSLM or AC voltmeter can handle the DC output or line voltage. If they cannot, simply use a capacitor (this replaces the high-pass filter) in series with the meter probe to take both readings. The same equation holds. Again, the various impedances would be plotted in the spreadsheet of impedance versus frequency.

$$Z_0 = \frac{V_b R}{V_a}$$

The DC power supply readings will be a smoother line, making it much easier to read directly off the spreadsheet graph. This higher output impedance could starve some circuits being powered by this supply. A line simulation network such as that depicted in Figure 4.6 can be developed using the value of the low ohms in series with the inductor read from the graph. Then these two series elements will be in parallel with the higher upper-frequency ohms.

Some power supply designers "cure" this problem with a capacitor tied across the output. This capacitor can oscillate with the inductance and cause ringing at the output of the supply. The best way is to calculate the maximum resistance that the following circuits can tolerate, say, 2 ohms. Draw the 2-ohm line across the same printed spreadsheet graph and read the frequency where the two lines cross. The first line is the 2-ohm line and the second is the inductive line from the graph, and say this cross point is 1,800 Hz. Divide this by 2.5, giving 720 Hz here. Then calculate the capacitor needed to equal 2 ohms at this frequency. The answer for this problem is 110 μF in series with the 2-ohm resistor tied across the power supply output line. If the curve was plotted again using the same setup, the curve would start in the same way and head up 45 degrees along the inductive direction to above 720 Hz, then head 45 degrees downward in the capacitive direction and flatten out on the 2-ohm line. A good source for this is the catalog *Power Conversion Design Guide and Catalog* (Calex Manufacturing Co., Pleasant Hill, CA 94532).

The output impedance of these supplies is important in EMI design because the wrong simulation impedance network could be used. Without the decoupling capacitor and resistor, the line simulation network looks like a small milliohm resistor in series with a small inductor. These two components are paralleled with a resistor value of the upper flattened-out resistance. With the decoupling network used, the output could be

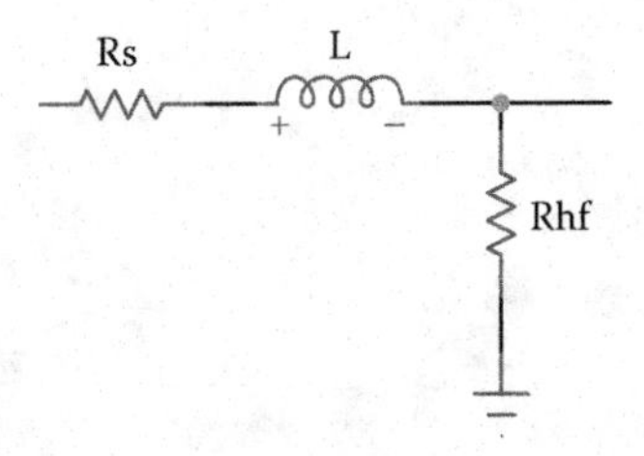

FIGURE 4.6 Line simulation network: a typical DC supply output impedance.

replaced with a fixed resistor, as in our case, using 2 ohms. The best way to measure the line impedance is still with a network analyzer, if you have access to one.

4.6 Other Source Impedances

A lightning strike will generate common-mode energy between the line, or lines, and ground. Assuming this happens at some distance down the line, the voltage divided by the current will be the characteristic impedance developed between the line and ground. This is the square root of the inductance divided by the capacitance between this line and ground. The inductance is around 1.5 μH per meter, and the capacitance per meter would be less than 1 pF. It should be obvious that this impedance will be quite high.

5

Various AC Load Impedances

The load impedance varies with different types of loads. Very few loads are truly resistive, as most are either capacitive or inductive. Some loads generate high-current pulses at twice the line frequency or twice the switching frequency, whereas others require high-frequency currents. All these statements, along with the rest of the information in this section, assume that none of these loads have any input filtering or any other circuitry to offset the conditions mentioned here.

5.1 The Resistive Load

The resistive load is the easiest for the EMI filter to handle. This assumes that the storage capacitor is left out in the circuit depicted in Figure 5.1. The only error would be the crossover error. This circuit can be with or without a transformer, and the results are the same. The power factor of the load is near unity in either of these cases, and the inductors of the EMI filter on the line side of Figure 5.1 must handle the normal RMS current. This means that the normal design method that most magnetic engineers use to design the inductors will give the desired results. These inductors will not saturate at the peak current. If the load does use the diodes, this will add noise problems to the already severe switcher noise from the circuitry represented by the resistor. The EMI filter must attenuate the entire load of the noise spectrum.

5.2 Off-Line Regulator with Capacitive Load

The most common circuit used today is the off-line regulator as shown in Figure 5.2. Again, this could be with or without the transformer. The voltage output of the diodes feeds a large storage capacitor C1, which in turn feeds a switcher.

The load impedance also varies from nearly short to open, depending on whether the rectifier diodes are turned on or off. This depends on the conduction angle, or instantaneous part of the sine wave voltage that is fed to the diodes. The diodes turn on if the sine wave voltage is greater than the capacitor stored voltage at that instant; in addition, we have the diode voltage drop and the *IR* losses. High current charges the storage capacitor

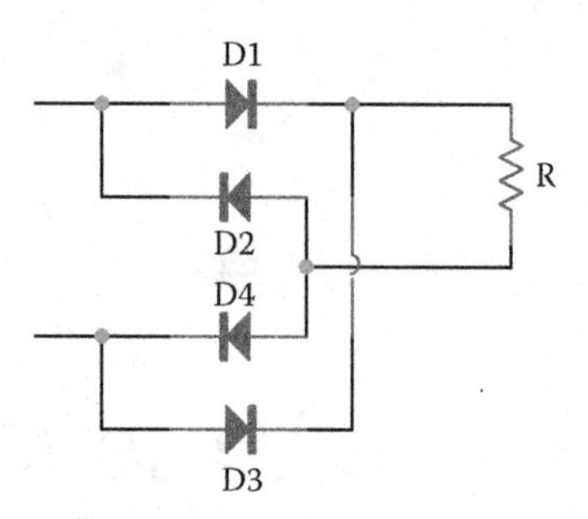

FIGURE 5.1 Diodes with load resistor.

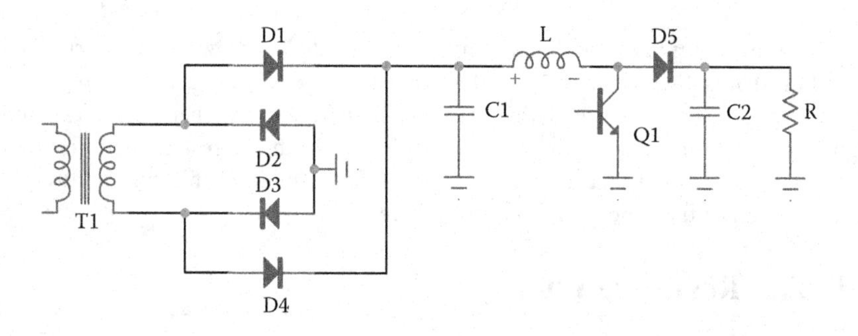

FIGURE 5.2 Power supply with capacitive load—C_1 storage capacitor.

during the turn-on period. The high-current pulses on the capacitor side of the diodes are often called sine wave pulses. But the curve shows that the main current pulse is well ahead of the 90-degree point (ahead of 1.57 radians on the plot), and the shape is not sinusoidal. On the diode side, this is rich in even-order harmonics, whereas the line side is rich in odd-order harmonics. Figure 5.3 starts at the angle of conduction of the diodes (in this case, 0.8 radians, and a few degrees later is the angle where the charge and discharge currents of the storage capacitor are equal. This is the lowest storage voltage point of the cycle (lowest ripple voltage) that feeds the load. Next follows the peak current angle that is well ahead of the 1.57-radian grid line (approximately 71.6 degrees). Past the 90-degree point, where the capacitor charge and discharge currents are again equal, is the maximum stored voltage point feeding the load (highest ripple voltage). The cutoff point (in this example, 2 radians) follows this. The curve repeats in the next half cycle. Just add 0.8 and 2 to π (the next conduction start angle) for the diode side, and the second half is negative for the line side. If the load requirement drops, the start angle increases while the peak load current also drops. The stop angle stays about the same and may even increase slightly.

The curve was developed using the following values: R_s = 1 ohm, R_l = 22 ohms, C = 0.001 farad, F = 60 Hz, E_m is the peak line voltage fed to the diodes, and N = 1 for full-wave rectification.

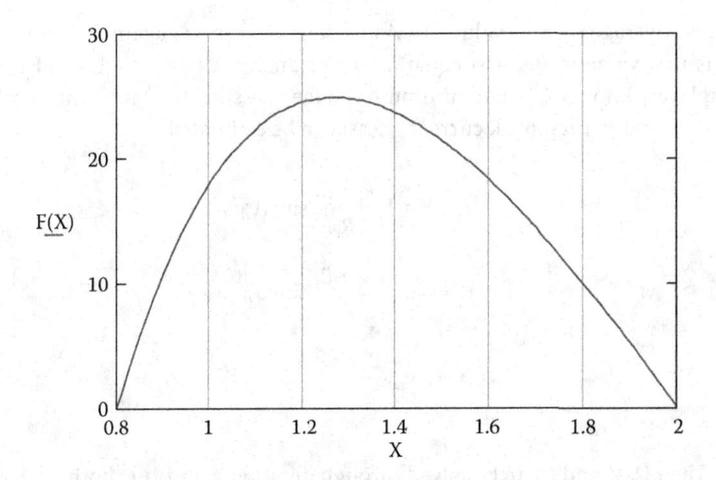

FIGURE 5.3 Diode current for a half cycle.

The question at this time may be why this is being discussed here. This is to show what the true peak transformer current is likely to be, so that it can be designed without saturating or overheating. The true RMS current is used for sizing the wire and to design the filter inductors. This really is the actual requirement for the EMI filter. The RMS current may be 8 A, while the peak current could be well above 25 A using this circuitry. The EMI filter must be designed to handle the peak current.

Figure 5.3 was developed by the following method:

Equations for capacitor rectifier circuits. In the equations, A is the start angle of conduction in radians where the diode is just turned on; B is the diode turn-off angle, also in radians; and N is 1 for full-wave and 2 for half-wave rectification.

$$\sin(A) = \sin(B)e^{\frac{(B-A-N\pi)}{\omega CR_L}} \tag{5.1}$$

where

$$Y = \pi - \tan^{-1}(\omega CR_L)X = \tan^{-1}\left(\frac{\omega CR_S R_L}{R_S + R_L}\right)$$

Equation 5.1 assumes the engineer knows the line frequency F, so $\omega = 2\pi F$ can be calculated, C is the storage capacitor in farads, R_S is the line resistance (typically 1 ohm), and R_L is the load resistance (usually the lowest value of resistance—highest current). The values of X and Y are then calculated along with $\tan(X)$. Substitute these into the first two equations. In one equation, guess a value for A and solve for B. Insert B into the second equation and solve for A. Average the two values for A and start the process again by substituting this new value in the first equation. Once angles A (start angle) and B (stop angle) are known, U_1 (first minimum voltage guess), V_1 (first maximum voltage guess), and P_1 (first peak current guess) can be estimated

$$U_1 = \sin^{-1}\left|\frac{R_s + R_1}{R_1}\sin(A)\right|$$

$$V_1 = \sin^{-1}\left|\frac{R_s + R_1}{R_1}\sin(B)\right| \tag{5.2}$$

$$P_1 = \frac{(A+B)}{2}$$

Then U, V, and P can be solved through iteration. Substitute both the known and estimated U_1, V_1, and P_1 into one term and solve for the other U, V, and P. Average the two and resolve.

$$\sin(Y+X-B) = \sin(Y+X-A)e^{\frac{(A-B)}{\tan X}}$$

$$\cos(X-U) = \frac{(R_S+R_L)\sin(Y+X-A)e^{\frac{(A-U)}{\tan X}}}{\sin(Y)}$$

$$\cos(X-U) = \frac{(R_S+R_L)\sin(Y+X-A)e^{\frac{(A-V)}{\tan(X)}}}{\sin(Y)} \tag{5.3}$$

$$\frac{\cos(Y+X-P)}{\cos(Y+X-A)} = e^{\frac{(A-P)}{\tan(X)}}$$

Then E_U, the minimum voltage at angle U (in radians), and E_V, the maximum voltage at angle V (again in radians), can be found.

$$E_U = \frac{E_M R_1 \sin(U)}{R_s + R_1} \qquad E_V = \frac{E_M R_1 \sin(V)}{R_s + R_1} \tag{5.4}$$

where E_M is the peak line input voltage. Also, the peak current at angle P can be calculated.

$$I_M = E_M \cos(X)\frac{\sin(P-Y-X) + \sin(Y+X-A)e^{\frac{(A-P)}{\tan(X)}}}{(R_S+R_L)\cos(Y)} \tag{5.5}$$

Then the equation of the current during conduction is

$$I = E_M \cos(X) \frac{\sin(\omega t - Y - X) + \sin(Y + X - A) e^{\frac{(A - \omega t)}{\tan(X)}}}{(R_S + R_L)\cos(Y)} \tag{5.6}$$

Equation (5.6) is plotted in Figure 5.3. It should be noted here that this value of ωt lies between A and B. The voltage during conduction is

$$E = \frac{E_M R_L \cos(X)}{(R_S + R_L)} \left[\sin(\omega t - X) - \frac{R_S}{R_L} \frac{\sin(Y + X - A) e^{\frac{(A - \omega t)}{\tan(X)}}}{\cos(Y)} \right] \tag{5.7}$$

However, the requirement for the proper design of the EMI filter is the value of I_M. This is the current the transformer must handle without saturating or overheating. Then:

$$B_{KG} = \frac{1.55 L_{\mu H} I_M 10^{-2}}{N A_C} \tag{5.8}$$

where the peak current is I_M, $L_{\mu H}$ is the inductance, N is the turns ratio, and A_C is the iron core cross-sectional area in square inches. This is presented here only to show what happens so that problems can be avoided. In other words, you do not have to solve any of these equations.

Power factor:

$$\begin{gathered} P_F = \frac{\text{true power}}{\text{apparent power}} = \frac{E_1 I_1 \cos(\phi)}{VA} \\ E_1 \approx V \qquad A = \sqrt{\Sigma(I_N)^2} \\ P_F = \frac{E_1 I_1 \cos(\phi)}{E_1 \sqrt{\Sigma(I_N)^2}} = \frac{I_1 \cos(\phi)}{\sqrt{\Sigma(I_N)^2}} \end{gathered} \tag{5.9}$$

Once the peak current angle (P in equation [5.2]) is known from the equations, subtract P from $\pi/2$ (90 degrees) to get φ. The current fundamental frequency's peak must coincide with the peak draw of I_M in equation (5.5). This (φ) is the angle of lead. To finish the power factor, use the RMS value of the fundamental current I_1 and the summation of the total current squared.

These approximate equations still hold true even if a transformer is inserted ahead of the off-line regulator. This current pulse is not truly a sine wave pulse, as most people describe it. These equations are included here because some engineers appear not to realize what happens when they increase the storage capacitor value. In general, we can say that as the storage capacitor increases in value:

1. The conduction angle decreases even though the peak-to-peak ripple voltage drops.
2. The peak current increases and moves farther away from 90 degrees.
3. Transformer and IR losses increase, which heats the transformer and filter inductors.
4. The wave shape is less sinusoidal.
5. The harmonic content increases.

The total size and weight are smaller with the critical inductor for all serious specifications, but might not hold for U.S. Federal Communications Commission specifications. Finally, these equations were developed so that the full 3rd, 9th, and 15th harmonic currents show why the neutral leg of the three-phase Y EMI filter heats. This is one reason why people ask, "Why is my return, or ground, filter much hotter than the other three?" The RMS value of the current is usually what the EMI filter designer and transformer designer are given to size the units. These designers are not given this high peak current needed to properly design their units. The question is, "What will this high-current pulse do to the filter inductors?" Most inductor designers design the inductor somewhere around half-flux density at the RMS value of the current. This is the wrong approach for the design of this AC EMI filter inductor because the inductor can saturate. The equation, for powder cores, is

$$H = \frac{0.4\pi N I_P}{M_{PL}} \tag{5.10}$$

where N is the number of turns, I_P is the peak current feeding the charging capacitor, M_{PL} is the magnetic path length in centimeters, and H is the magnetizing force. The maximum flux density, B_m, is a constant for the core material. The relationship between B and H is the permeability, μ. As H increases due to the large current pulse, the permeability drops. The permeability μ is a key player in determining the inductance, L. The core material is spoken of as "soft" because of the BH curve. The BH curve is S-shaped, or sigmoid, and is not made using square loop material. These cores require a strong magnetizing force, H, to drive the core into saturation, but still the permeability drops. A hard core, or square loop core, is driven quickly into saturation. The soft core is the type of core material needed for EMI filter inductors. Cores that have square loop characteristics are gapped, if they are to be used at all, thereby reducing the hard magnetic core to a softer core. Soft cores are also often gapped to make them even softer (less sigmoid) and for DC applications. The equation to calculate the inductance for powder cores is

$$L = \frac{0.4\pi N^2 A_C 10^{-2}}{g} \; H \tag{5.11}$$

where the only new term is the cross-sectional core area, A_C, in square centimeters. So as the value of μ drops, the value of L drops, and so does the inductive reactance. All inductors must be gapped, but powder cores have distributed gaps. These are soft cores, and equation (5.11) works for powder cores. For these cores that require gapping, such as C-cores and tape-wound toroids, we can calculate the inductance in terms of REL.

$|M_{PL}/\mu|$ is the core REL. The gap REL is the gap/μ(air). The gap, g, is much greater than the core REL, so equation (5.11) becomes

$$L = \frac{0.4\pi N^2 A_C 10^{-2}}{g} \; \mu H \tag{5.12}$$

What happens if the inductor saturates during this current pulse peak? The answer is that switching, diode, and other noise can ride through the saturated filter inductor during this time. These noise spikes can therefore show up on the peak of the AC sine wave voltage. A scope should show noise at the wave peak if this happens. Another way to avoid this is to design an inductor using a gapped core. This tilts the *BH* curve, therefore requiring a much higher magnetizing force to drive the inductor into saturation. It also makes the permeability much lower but more constant. The point is that the EMI filter will be bigger, it will weigh more, and it will cost more when using the off-line regulator because of the high-current pulses.

5.3 Off-Line Regulator with an Inductor ahead of the Storage Capacitor

Adding an inductor ahead of the power supply filter, or storage, capacitor widens the current pulse width and lowers the peak current pulse. See Figure 5.4. If the value of the inductor is equal to or greater than the critical inductance, the current flows all the time, and the current through the inductor is the average current. The normal inductance design method will work in this application, and the critical inductance is

$$L_C = \frac{R_O}{6\pi F} \tag{5.13}$$

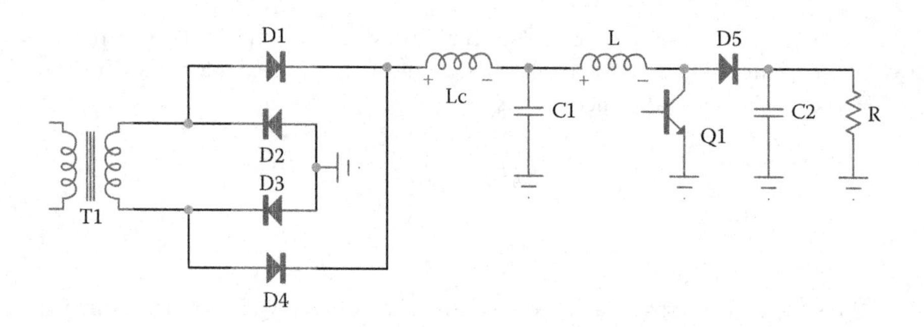

FIGURE 5.4 Off-line regulator with critical inductance, L_c.

where R_0 is the load resistance calculated using the lowest, worst-case current and the highest line voltage, and F is the line frequency. The disadvantage is that the stored voltage across the storage capacitor drops from the peak value of the AC voltage to the average voltage. This also adds to the weight of the power supply but cuts down on the EMI filter weight because it is not necessary to handle this high peak pulse current.

Some power supply company designers have been very proud of their light, compact power supplies. The power supply designers would ask the company to design a filter for this small power supply, but they would not understand why the filter is larger than their supply and weighs more. To keep their power supplies this small and light, they have eliminated all the normal internal EMI protection. Therefore, the output noise is high and, ultimately, the EMI filter is quite large.

5.4 Power Factor Correction Circuit

Today, power factor correction circuits are an essential aspect of any piece of equipment that has an off-line power supply (Figure 5.5). Engineers have reported power factors as low as 0.4 in days gone by, resulting from the power factor of the power supply and the EMI filter because of the larger capacitors. Power supplies that use off-line regulators have power factors that are now well above 0.9, and this is with the EMI filter. The reason for this is because the transformer and EMI filter have lower current that is in phase with the voltage. The power factor correction circuit works by switching the diode output voltage without initially storing the energy in a large filter capacitor. There is a small capacitor in Figure 5.5, denoted C_X. This capacitor is typically about 0.2 μF and is used to store a small voltage when the sine wave crosses the axis at 0 and 180 degrees.

These are switched at 80 kHz or higher frequency for most applications. Thus, the current is in phase with the line voltage. The power factor switcher generates the power factor back to near unity. This means that the output impedance of the EMI filter must be very low compared with the conducting load of the power factor correction circuit switcher frequency. The inductor conducts current in the same direction when the switch is open and closed, so the current through the diodes is a sine wave with a small triangular wave superimposed on top of it at the switcher frequency. It is the job of the EMI filter to attenuate this switcher frequency without starving the switcher. This is

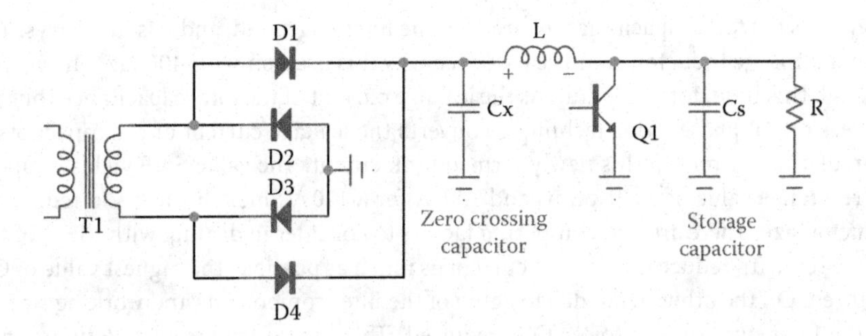

FIGURE 5.5 Power factor correction circuit with zero crossing capacitor.

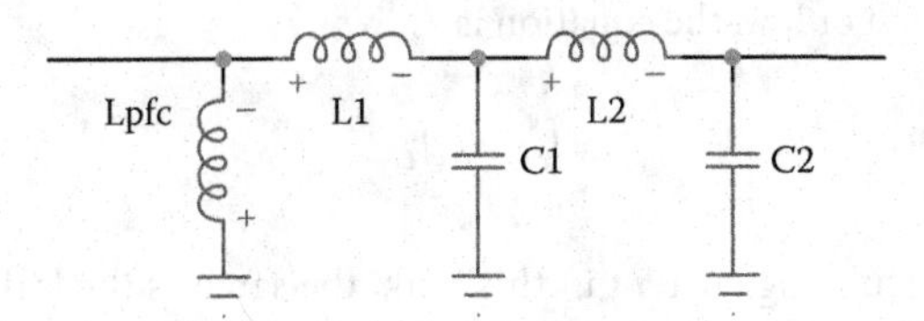

FIGURE 5.6 Power factor inductor.

another reason why a filter designed for very similar specifications may work well for one group and fail for another. The first group could be using an off-line regulator, and the second group a power factor correction circuit, and the output impedance of the filter may be inductive (such as a T. Filter). The inductive reactance would be low to the harmonics to the off-line regulator but high to the power factor correction circuit. This would starve the power factor correction circuit and may create heavy runaway oscillations. The disadvantage of initial power factor correction circuits was that they were not close to 100% efficient. One of the main reasons for demanding power factor correction circuits was to allow more devices to be plugged into the wall sockets. These people did not account for the lower efficiencies of these circuits.

There is also a power factor correction coil (Figure 5.6). In most designs, the EMI filter looks very capacitive at the power line frequency, and some specifications demand a near-unity power factor for the filter for two reasons: obviously, for power factor correction, and also for leakage current problems. This technology is archaic, at best, and is mainly seen on 400-Hz power lines. At the line frequency, the impedance of the inductors in the EMI filter is very low, so the capacitors add in parallel to a value, in Figure 5.6, of 2*C*. The inductive reactance of the power factor correction coil must be equal to the total capacitive reactance at the line frequency. This returns the power factor to near unity and the leakage current is reduced. If C1 and C2 are both equal to 3 μF, then equation (5.14) will yield the following:

$$L = \frac{1}{\omega^2 [C_1 + C_2]} \cong 26\,mH \tag{5.14}$$

where C is the total capacitance in farads of the filter to ground, and L is in Henrys. This is a rather large inductor, so it is easy to see why it is used only for 400 Hz. The current through this inductor is the same as the leakage current of the total capacitance (but 180 degrees out of phase). This technique converts the leakage current of the capacitors to a circulating current in this newly formed tank circuit. The value 6 μF yields a capacitor reactance value of 66.31 ohms and 1.81 A for a 120-V line. This helps to reduce the inductor size. There are two conflicting factors to consider in dealing with the Q of this tank circuit. To reduce the leakage current as much as possible, the highest value of Q is required. On the other hand, due to aging of the filter components and working stresses on the filter over time, a lower Q is required. This is also true in installations where the power line frequency drifts over a wide range, such as remote power generators. This frequency drift would cause the network to be off-tuned from the center frequency and be operating on the side skirts of the impedance curve of the parallel tank circuit. Whatever the Q, high or low, the equation is

$$I_{circ} = QI_{Line} \tag{5.15}$$

where I_{circ} is the circulating current in this tank, the same as the leakage current prior to addition of the power factor correction coil, and I_{Line} is the new leakage current. The reason for the concern about leakage current is safety for anyone touching the filter, or the unit that the filter is mounted in, if the safety ground has been removed or broken off. The electric shock could be lethal when this ground is cut. Equation (5.15) shows that the leakage current has been reduced by a factor of Q.

RFI Corp., on Long Island, tunes the power factor correction coil for each filter before shipping. This technique allows them to ship these units as matched pairs with higher Q values. Otherwise, the Q should be limited to about 10. RFI attaches these power factor correction coils on the load side, and often in add-on doghouses. Others mount these power factor correction coils at the front-end on the line side. We know of no functional electrical difference.

The specifications are leaning toward measuring the actual ground current rather than specifying the capacitance to ground. Using this trick would allow larger capacitors to ground and enhance the common-mode action. Using a Q value above 10, the 0.02 μF for 400 Hz could be changed to 0.2 μF. This would reduce the value and size of the common-mode inductor, here called the Z (Zorro). This would enhance the common-mode attenuation so that the filter easily passes the insertion loss requirement while reducing the value of the Zorro inductor. The only concern here is that the components and the line frequency must be stable to ensure that the units are not detuned, or the leakage current will rise. This also returns the power factor back to near unity. The power factor correction coil becomes very large in size for 50 and 60 Hz and is rarely used.

5.5 Transformer Load

If the filter designer knows that the load is a transformer, knowledge of the transformer is necessary. Try to get the customer or the transformer manufacturer to tell you the transformer primary inductance. The reason is that the total filter inductance in series across the filter must be much lower than the primary inductance. Otherwise, the filter and transformer inductance form an inductive voltage divider.

Typically, the primary inductance is well into the millihenry range, possibly 50 mH, or more. For a power transformer, this is not a problem, but it should be checked. Where a potential problem may show up is with autotransformers and multiphase transformers. Autotransformers typically have much lower primary inductance that often causes this problem. The autotransformer is smaller than the isolation transformer, and this allows the smaller primary inductance. The total EMI filter inductance should be less than 2% of the primary inductance to reduce the voltage divider effect.

5.6 UPS Load

Another load to consider is the UPS (uninterruptible power supply) load. At least in the past, the typical UPS had zero crossing spikes that challenged the EMI filter. These spikes are very high frequencies and require quality capacitors of the feed-through type giving very low impedances to the 10th harmonic of the fundamental spike frequency. These spikes are typically at 25 kHz and higher, requiring a large capacitor with SRF to well above 250 kHz. Power specialists that have analyzed power system problems concluded that a UPS was needed, only to find out that a UPS system was already installed and was creating the problem. Capacitors with long leads or connections cannot be used here, especially for the final capacitor facing the UPS because of the low SRF caused by the higher equivalent series inductance (ESL) and equivalent series resistance (ESR) of the leads. Often, a feed-through capacitor of 3.5 μF is enhanced with a leaded capacitor, sometimes referred to as a *hang on*, and is used to bring up the total capacitance value. The leads must be very short, and the two capacitors can form a parallel resonance circuit that can defeat the EMI filter. This occurs when the larger capacitor becomes inductive and then resonates with the smaller capacitor.

To sum up this section: Attempt to find out as much as possible about the load. Is the load a power supply and, if so, what type? If the power supply has a power factor correction circuit or an inductive input, the EMI filter must have low output impedance. The same is true if the power supply has a capacitive input; if so, the inductors of the EMI filter must be designed to handle high-current pulses. If the filter feeds a transformer, check the primary inductance and make sure the total EMI filter inductance is less than 2% of this primary inductance.

6

DC Circuit—Load and Source

There is a key difference between AC and DC circuit designs for EMI filters. It is understood that many off-the-shelf electromagnetic interference (EMI) filter manufacturers sell filters specified to work for both alternating and direct current. This adds to the difficulty in design because the sources are very different. Filter capacitors must be designed to handle the AC voltage and the total harmonic current, and this calls for larger margins at 400 Hz. AC capacitors are typically designed to handle 4.2 times the RMS working voltage, whereas DC capacitors must handle 2.5 times the working voltage.

Let's assume a line with 250 V specified to ground. If the system is AC, this equates to a 1,050-V margin, which means that the initial high potential, or hipot, will be 1,800 V. This would increase the margins of the capacitor for the higher AC voltage. The inductor should be gapped to handle the direct current; otherwise, the inductor will easily be driven into saturation. This dual requirement adds to the overall cost of the filter. In the preceding example, the DC 250-V line requires 625 V, calling for an initial hipot of 1,000 V. The filter manufacturer helps to cut the cost of these filters by building sizable in-house orders for shelf stock. These filters are usually grouped in families of different current values with the same specifications such as leakage current and dB attenuation. The same parts are used for all the different current values when possible, and this drives down the component costs, which ultimately makes the parts less expensive and reduces the cost to the customer. The filter customer can cut costs by having more than one application requiring the same filter. This increase in the number of filters required over a period of time cuts the filter price. This is true for all filter types.

The loads are also different. Some are resistive, as in heaters that maintain mountaintop repeaters above a certain temperature. Some equipment may use this supply voltage directly; others may be switchers, for which the inductors should be gapped for pot cores or C cores. MPP (molybdenum permalloy powder) and powder cores are already gapped as a result of the manufacturing process. Gaps for the MPP, Hi Flux (HF), powdered iron, ferrite, and now the new cores available have distributed gaps throughout the core.

This chapter discusses recommendations for elements within the circuit that reduce the EMI so that the insertion loss requirement of the filter is reduced, thereby reducing the size, weight, and cost of the filter. Other techniques are also discussed.

6.1 Various Source Impedance

The DC power line feed is typically very short, providing link coupling well up into the megahertz range. Often, it uses the normal chassis ground, making a balanced circuit impossible. This is where tubular filters play a role using feed-through capacitors without line-to-line capacitors. Some have capacitance only, others make up the L and π structures, and the rest are T filters. This DC power is the type found in aircraft, shipboard, telephone company, and mountain repeater sites. This power is furnished by various arranged systems such as battery racks, standby generators, solar panels, diesel generators, and wind generators. The battery feeds stored energy to the various systems and helps to regulate the voltage. These large batteries are called deep-cycle batteries. Deep-cycle batteries have more clearance between the inside battery bottom and the bottom of the battery plates. This allows a deeper discharge because of the increased clearance room for plate material and debris. These can feed the entire system from several hours to many days, depending on the system. They are designed in this way where the conditions warrant the higher costs or are referred to as *life threatening*; the highest cost systems require more standby batteries, generator fuel storage, etc. These types of conditions can be caused by any outside power failures, such as from a solar panel on a series of cloudy days, downed power lines, fuel supply outages, and bad weather conditions. The plates of the battery rack act as a capacitor and shunt the middle frequencies of the unwanted conducted emission noises to ground. The high frequencies are not attenuated because of the inductance of the cable feed and the battery plates. The radio frequency (RF) current on the power lead is radiated as an *H* field, and this is what should be avoided by filtering. The DC output normally feeds switching converters, which create most of the RF noise along with the diodes following the switcher and circuit parasitic oscillations created by high-frequency power switching action, etc.

The other type of DC system is from an AC power supply, and again the feed is very short. This DC normally feeds a switching converter. The difference is that this output impedance is high at the switcher frequency even though the output impedance is only a few milliohms from DC to 10 Hz or so. The output impedance is inductive above this point and rolls off flat or resistive at, for example, 5 kHz. This power supply looks like a milliohm resistor in series with an inductance, and this is shunted with the higher resistor value. If left unchecked, this situation starves the switcher because the inductive reactance of the inductor at the switcher frequency will be high impedance equal to R_{hf} (Figure 6.1). Is this true for all DC power supplies? No, in some rare cases the customer informs the designer long before the design is complete of such specifications, e.g., the power supply must have low output impedance at a specific high frequency, including other conditions that the supply must provide. The power supply designers are able to make the output impedance low at switcher frequencies around 100 kHz. Does the filter designer need to know this output impedance of the DC supply feeding the filter before designing the new filter for the remote switcher?

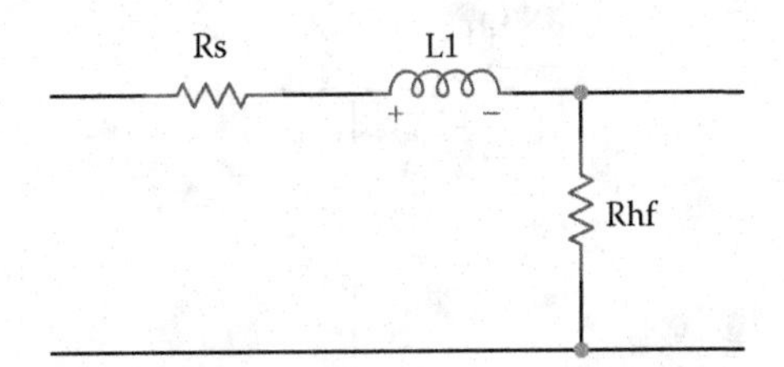

FIGURE 6.1 Line simulating network.

What happens if the output impedance of the DC supply is low without the filter? The switcher will not be starved, but the high RF current pulses will radiate, thereby producing a very high *H* field. What happens if the output impedance of the DC supply is high without the filter? It has been stated repeatedly in this book that the switcher will then be starved, but the weak RF current pulses will produce a very weak *H* field. The point here is that the filter "fixes" either condition, so the filter designer is not really concerned about the output impedance of the power supply. This is relatively straightforward to design and is discussed in the next section.

6.2 Switcher Load

Most loads today are of the switching type. The output impedance of the filter at the load or switcher end must be such as not to starve the load. This assumes that the switcher does not have a capacitor at the input to lower this impedance. The capacitors close to the switch are classed as part of the filter. If this capacitor is in the circuit without the filter designer's knowledge, it could detune the output network, thereby lowering the cutoff frequency of this filter. The capacitor values add because they are in parallel. It is also assumed that the connecting wire is not a long lead, as this would appear as an inductance tending to split the capacitors. In parallel, these two capacitors can form a parallel tank circuit when the larger capacitor is above its self-resonant frequency (SRF).

This DC application would require a single L filter so that the capacitor facing the load is as high as possible (Figure 6.2). This is to reduce the drop in the DC voltage while the switching converter is in conduction and drawing current from the source. Typically, the capacitor is sized to ensure that the energy demanded by the load during the conduction period is supported by the capacitor. This also assumes that the voltage droop on the input is within defined limits. The capacitor may be sized using a simple approximation as follows:

$$V_C(t) = \frac{1}{C}\int_0^T i(t)dt \Rightarrow C = \frac{i(T - t_0)}{V_C - v} \tag{6.1}$$

where V_C is the nominal capacitor voltage, v is the droop voltage, and $T - t_0$ is the conduction period. So, the capacitor will provide the energy required during the conduction period. If double L filters are used, the value of the final capacitor is smaller than half the original. This could allow the voltage feeding the load to drop more than that for the

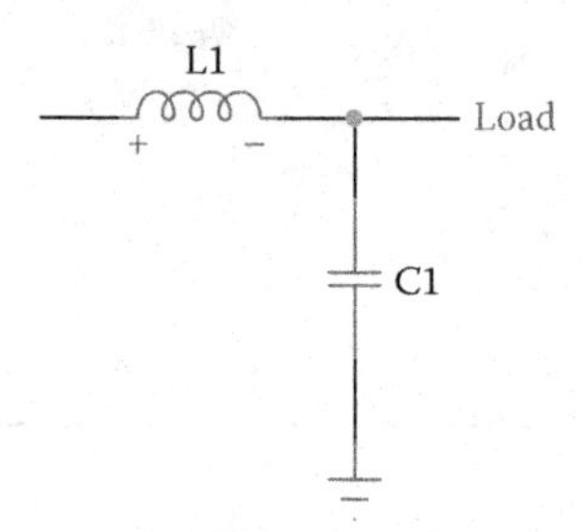

FIGURE 6.2 Single L filter: capacitor at load.

single L filter. This is still true with the double L, even though the ripple voltage on the feed wire is the same. If the return is through the chassis, the capacitor facing the load should be a feed-through type. Otherwise, it should be a line-to-line capacitor in case the system has a return power lead. The input inductor of the filter has high impedance and reduces the *H* field no matter what the source impedance of the power supply happens to be. These are easy to design and the method is as follows.

For example, the switcher frequency is 60 kHz, the DC voltage is 60 V, and the *peak on current* is 1.2 A. Divide the voltage by the switcher "on" current to find the "on" impedance of the switcher, and divide this "on" impedance by 10 to minimize the voltage drop. Define the capacitor value by making the capacitive reactance equal to this impedance at the fundamental switcher frequency, 60 kHz. For the inductor, multiply the "on" impedance by 10 and then solve for the inductor reactance equal to this impedance at, as before, 60 kHz.

$$\frac{V_{dc}}{\hat{i}(10)} = R_C = 5\Omega \rightarrow C = \frac{1}{\omega R_C} = 0.53\,\mu\text{F} \tag{6.2}$$

$$\frac{V_{dc}(10)}{\hat{i}} = R_L = 500\Omega \rightarrow L = \frac{R_L}{\omega} = 1.3\,\text{mH} \tag{6.3}$$

This is rather a large inductor, but it should be easy to design because the main current is DC. If the duty cycle is 50%, the circuit carries only 0.6-A DC with a small ΔI AC current. However, if the circuit is balanced, then two inductors of 670 μH each are needed and must meet the same current requirement. The small capacitor of approximately 0.6 μF is wired between the two inductors and faces the load. To calculate the actual inductor current, the pulse width is needed. Every pulse in this example is 4-μs duration multiplied by two for the two pulses per cycle, and the 60-kHz cycle is 16.66 μs. Therefore, the average DC current through the inductor(s) is

$$\frac{2\hat{i}T_{on}}{T_{total}} = \frac{2\times 1.2\times 4}{16.66} = 0.57\,\text{A} \tag{6.4}$$

The 1.3-mH inductor would have to carry only about 0.57-A DC, making this possible even if the filter is unbalanced. This can be made into a simple formula:

$$C = \frac{10\hat{i}}{2\pi F_{sw} V_{dc}} = \frac{10 \times 1.2}{2\pi 60{,}000 \times 60} = 5.305 \times 10^{-7} = 0.5305\mu F$$
$$L = \frac{10V_{dc}}{2\pi F_{sw}\hat{i}} = \frac{10 \times 60}{2\pi 60{,}000 \times 1.2} = 1.326 \times 10^{-3} = 1.326mH \quad (6.5)$$

The equations give inductance in henries and capacitance in farads.

6.3 DC Circuit for EMI Solutions or Recommendations

This section is included here because most problem solutions deal with DC power feeds or are part of the power supply. Some of these may conflict, but it is important to use what works.

Some systems are developed piecemeal, or bottom-up, as follows. The initial concept or basic idea is based on a customer request, product application, or new technology. The various concepts are listed, followed by choosing a method of implementation. The various design solutions are selected, and the relevant circuits or circuit boards are combined to test the system. This is followed by system design iteration and mechanical adjustment to reduce weight and, thereafter, more circuit corrections to get the CPU (central processing unit) card into most of the available volume. Next, we need a power supply and EMI filter. Now, where can we place these? We don't have too much room for those. Are they really that important? We will tuck them in somehow where we have a small volume of 2.25 × 1.6 × 0.75 inches. We could place the power supply at one end and the EMI filter can go into that small volume. Does this sound familiar? Indeed it does. The system was designed by someone who does not understand two most important functions in any electronic system: the power supply and the EMI filter.

Usually, as described here, the power supply and EMI filter is often a last thought, and especially so when the project does not have power electronics engineers driving the design. People often wonder why the filter is so big and bulky and why it cannot fit in that little leftover volume next to the input transformer. The plan for power management and EMI should begin almost at program inception. If this is done, the EMI filter can be designed correctly and has a high probability of meeting the EMC requirements. With this in place, the weight and volume of the EMI filter will be what it needs to be. End of story.

6.4 Some Ideas for the Initial Power Supply

Include an inductor of critical value in front of the storage capacitor to remove the high current peaks. Lack of this inductor increases the size of the EMI filter; it weighs more and it costs more, especially for the military, whose specifications are harder to meet. Isolate the off-line regulator to reduce the common-mode noise and remember that a capacitor is not a capacitor at all frequencies. They all have a self-resonant frequency (SRF). Power supply storage capacitors of larger size are inductive by 80 kHz. Parallel them with a good-quality extended foil or ceramic capacitor, but watch for a resonant rise. Use the proper snubber circuits in the switching converters to remove voltage spikes due to parasitic inductance. Where possible, use soft switching diodes to reduce diode switching noise. Keep the various supplies isolated and use twisted pairs to power remote units rather than a ground return. Filter the isolated leads or closely spaced traces on the printed circuit board at the device end. In this way, the primary current in the traces or twisted pair is direct current rather than pulses at the switcher frequency. Lay out the power supply transformer input or switcher(s) to minimize magnetic coupling between the transformer and other susceptible receptors by increasing the distance between these devices or finding the best orientation of the transformer.

Sometimes the magnetic field radiating from or around the transformer is stronger in one direction than another—the lines of magnetic force may extend farther. Go to a toroid transformer, which is known to be "quieter" magnetically. Use single-ended converters rather than a flyback transformer, and do not use switchers known to be noisy, such as SCRs (this should be obvious). Place sections known to be noisy within a container or shield, which helps to stop the *H* and *E* fields. Cold-rolled steel must be quite thick to reduce the magnetic field generated by a transformer or any other device. These containers are often silver plated inside and out to enhance the surface conduction. The better the conduction, the better the *H* field is attenuated and the lower the current on the outer skin. This may require a *mu* metal can or foil, which would be thinner and lighter for the same attenuation of the *H* field but would cost much more.

Moves to increase the self-resonant frequency of the inductor or transformer of the switcher also reduce the parasitic or at least raise the frequency of the parasitic. This parasitic frequency often feeds all the way back to the filter, where it is attenuated. It can be attenuated with a smaller value of capacitance at the interface of the power supply and switcher. Filter these as close to their source as possible.

6.5 Other Parts of the System

Clocks and other sources generate more than their share of noise. Sometimes the noise does not start or originate in the clock, but is created either in the power supply or in other circuits that the power supply feeds. Make sure that each feed of the supply is filtered. Filter each feed such as the clock, DAC, or other device as close to the device as possible. Separate the ground systems for digital and analog. Separate grounds for the noisier systems so that the noise current on the ground lead does not induce a voltage in the quieter circuits which may be amplified. Shield the noise generators so that magnetic radiation of the *H* field is a minimum.

Keep the noise within the EMI filter inductors and other inductors in nearby areas, even though the cores are toroids (which are quieter and radiate less), by mounting these inductors in quadrature to reduce their coupling. This is especially true if

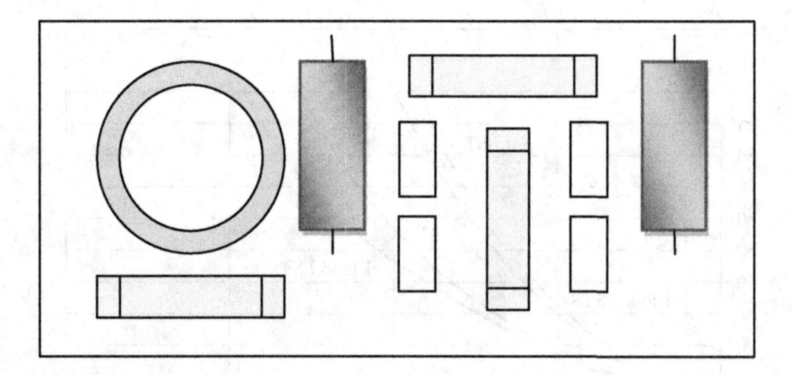

FIGURE 6.3 Toroids in quadrature.

the cores are not toroids, because they are known to have more flux outside the core material. This flux can induce currents into other susceptible devices that are in close proximity.

Figure 6.3 shows a balanced double "L," where the first two inductors feed the capacitor using the vee technique. The last two inductors in quadrature also use the vee technique. The lower left inductor could be magnetically coupled to the upper right inductor, but the increased distance and the capacitor between them reduce this tendency.

6.6 Lossy Components

Another technique is to use lossy components. One such technique is derived from transmission lines, where high frequencies are absorbed within the dielectric. Coaxial cable developers continue to research to find dielectrics with lower and lower losses, whereas others have used this phenomenon to enhance the losses in an effort to make lossy systems. One such company in Long Island, New York, is Capcon, Inc. Their material is tubular and can be threaded with the filter hookup wire; it provides substantial losses above 10 MHz and well into the gigahertz region. This depends on the total length of the material within the filter. Various internal diameters are available, allowing easy threading for capacitor, inductor, and other hookup leads. It has been shown that Capcon material functions better than most ferrite beads. The lossy component mentioned here does not fail either of these requirements. One foot of this lossy line material gives as much as 100 dB at 10 GHz. This material is also available in sheets with various thicknesses for enclosures and is very dissipative. Radiated energy can be reduced by covering the filter enclosure or a noisy device with this material.

Figure 6.4 shows the approximate loss of lossy suppressant tubing shielded (LSTS) and LST, which is the same tubing without shielding, measured per MIL-STD-220A.

Ferrite beads are sometimes chosen incorrectly. Determine the frequencies requiring additional loss and search for a ferrite bead that is in its resistive mode. This way,

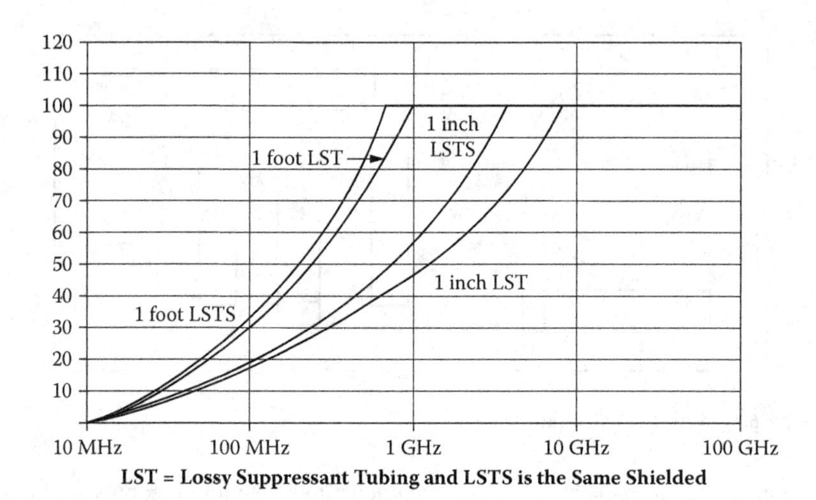

FIGURE 6.4 Capcon tubing and sheets.

the ferrite bead dissipates these unwanted frequencies rather than reflecting them back to their source, as this can add problems rather than fix them. In the past, many ferrite beads would saturate at around 5 A; however, this has improved over the years.

6.7 Radiated Emissions

The subject of this book is the design of the filter for conducted emissions. It is generally true for radiated emissions that if the conducted emissions are reduced, the radiated emissions are also often reduced. If the conducted emissions are well within defined limits and meet EMC directives, there is often little left to radiate. The opposite condition is also valid. If the conducted emissions are not eliminated, the device or equipment will almost certainly radiate. A condition that makes this statement erroneous is that the outer skin of the device is not a good conductor—plastic for the purposes of discussion. Although the conducted emissions may have been greatly reduced, the radiated emission will be almost as offensive. If the case or enclosure has to be plastic, there are conductive sprays that the enclosure can be coated with that help to reduce the emissions.

7

Typical EMI Filters—Pros and Cons

Many different types of EMI filter structures are used today. These are primarily the π, T, and L, with Cauer and RC shunts. They include double and triple filters and sometimes even quadruple uses of the mentioned types of filters. Each EMI filter has some positive and negative attributes. Each filter type lends well in certain applications, and in others the filter may not perform as well as required. This chapter presents the various filter types and discusses them in terms of structure, performance, and their individual merits. In addition, the transformer—if used as part of a filter structure—will add to the filter loss. This is discussed in chapter 11.

7.1 The π Filter

We start with a discussion of the pi (π) filter (Figure 7.1). This filter looks very good under the 220-A 50-ohm test specification. The line-side capacitor will work into the 50-ohm line impedance at the low frequencies. If this is the only test requirement specified by the customer, then the π will test well with this test technique. The π filter will pass with flying colors when considering the source. This is especially true for the three-phase type, for which the specification states, "Measure one phase with the other two phases tied to ground." The input and output capacitors of each π section are then doubled in value. This makes it easier for the filter designer to meet the specified loss within a given weight and volume. It means that the filter can easily pass the attenuation requirement with smaller values, package size, and weight.

In Figure 7.1, it is shown that the center capacitor is twice the size of the source and load capacitors. All inboard capacitors are typically twice the value of the end capacitors for all multiple π filters. On the other hand, this type of filter does not do well under the MIL-STD-461, naval, or the current-injection probe (CIP) tests. This is where a 10-μF capacitor shunts line to ground and increases the current through the output current probe and shunts the line-side capacitor (Figure 7.2).

The reason is that a filter capacitor facing the 10 μF is out of the circuit unless the capacitor is an order of magnitude greater than the 10-μF capacitor. This value would be much too high for good filter design. Large capacitors have very low self-resonant

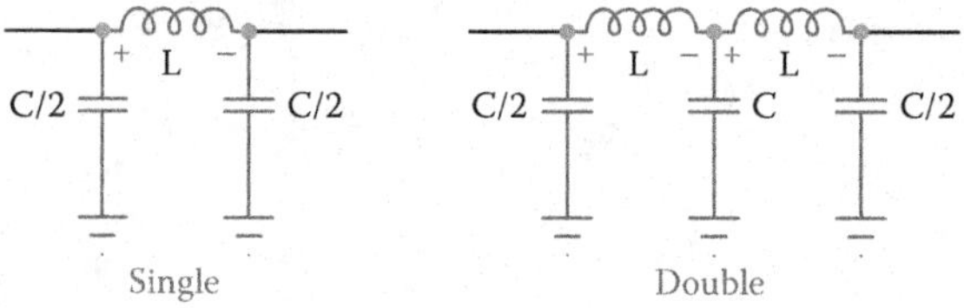

FIGURE 7.1 The π filter.

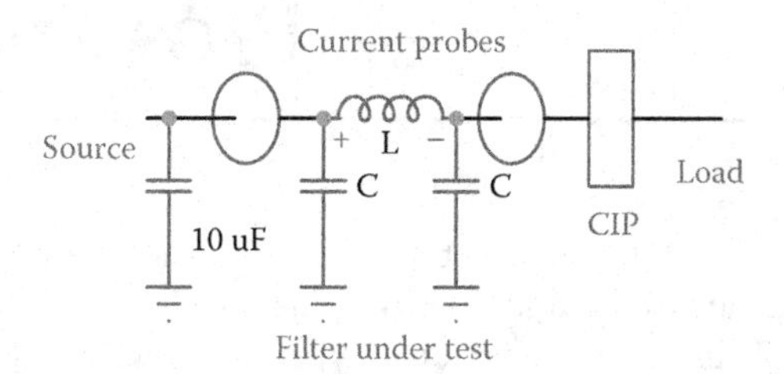

FIGURE 7.2 CIP π problem.

frequency (SRF) and would have to be shunted with smaller capacitors of good quality to make up for this low SRF. The receiver or spectrum analyzer is switched back and forth between the two current probes to determine the filter loss.

The three-element π looks like a two-element L filter at the lower frequencies in the MIL-STD-461 and naval tests. The loss is about 6 dB per octave for each element, so the 18-dB π filter is now a 12-dB loss per octave filter; 18 dB is for the single π, and the double π is 30 dB, for example. The π filter will also function well in some DC systems if the power supply switching frequency is high enough so that the capacitor impedance facing the load is small enough not to starve the supply. In addition, the capacitor impedance should not cause excessive voltage drop. The π filter is easily balanced by placing only half of the inductor needed in the high line and the other half in the neutral line. This changes the filter structure to a balanced configuration. The π filter may not always pass testing, as the filter loses the effectiveness of the front, or line-side, capacitor attenuation at the lower frequencies. A question may arise, "Why is the filter not doing the job when it properly passed the tests at the EMI test laboratory?" The real world is not reflected by the 220-A or 461 specifications, but the 461 specification is closer to reality.

7.2 The T Filter

The T filter (Figure 7.3) provides 18 dB per octave loss, and the double T gives 30 dB per octave (6 dB per element) loss. T filters work best in low-impedance lines (high current requiring small inductors). The line impedance is very low up to at least 100 kHz, but the MIL-STD-461 loss specifications typically start between 10 to 20 kHz. The inductive

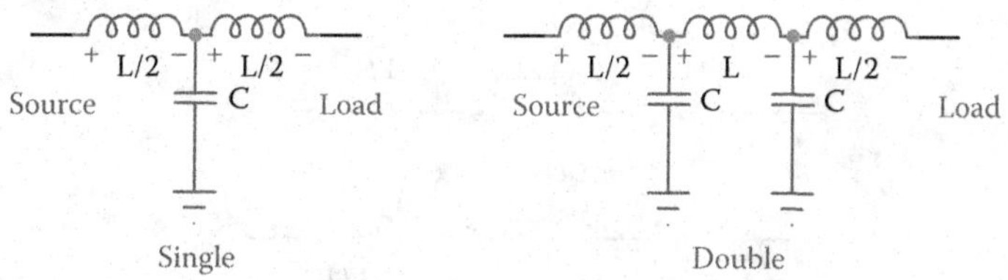

FIGURE 7.3 T filter.

input impedance of the T adds to the low line impedance. This gives the capacitor an input impedance to work into. These are also best in the higher current loads if the design method does not call for overly high values of these T inductors. This could result in the voltages soaring or dropping while feeding the load. It is not uncommon to see 115-V AC 60-Hz lines as high as 132 V feeding a light load requirement. This happens because the resonant rise occurs at a very low frequency and is usually caused by higher inductive values.

The T should never be used in the DC system if the load utilizes any PWM-switching power converters because the high impedance of the output inductor facing the load will starve the switchers. The switcher designer may have taken this into account by lowering the impedance with a capacitor at the switch input. This really makes the filter into a maladjusted double L, because the capacitor shunts the inductor facing the load, but this is less troublesome in DC than AC.

Note that the central inductor is typically twice the size of either of the two end inductors. The T filter can be balanced by removing half of the inductor's values and placing this half in the neutral leg, forming an H pad as shown in Figure 7.4. For all these filters that are balanced, it is necessary to be careful of hidden grounds within the system that can change the circuit configuration. As shown in Figure 7.4, the ground structure causes the bottom half of the circuit to drop out. Actually, it is worse than that, as the two capacitors are now in series with two ground inductors, thereby creating a filter structure that will be prone to instability and frequency magnitude slope error.

7.3 The L Filter

The L type is the filter most often used. The π and the T comprise three elements (or more if multiples are used) and should give 18 dB of loss per octave. The L filter, being only two elements, provides 12 dB of loss per octave. All of these loss figures refer to the loss starting above the cutoff frequency. A single L filter works best in the DC mode if the load has switchers. The inductor faces the DC source, and the capacitor (of high quality with a high SRF) would provide a low impedance path for the switcher frequency (Figure 7.5).

Why not a double L? The two inductors required for the same amount of loss would total less than the single inductor, and the same is true for the capacitor values. A smaller output capacitor may not furnish the needed energy storage for the following switcher.

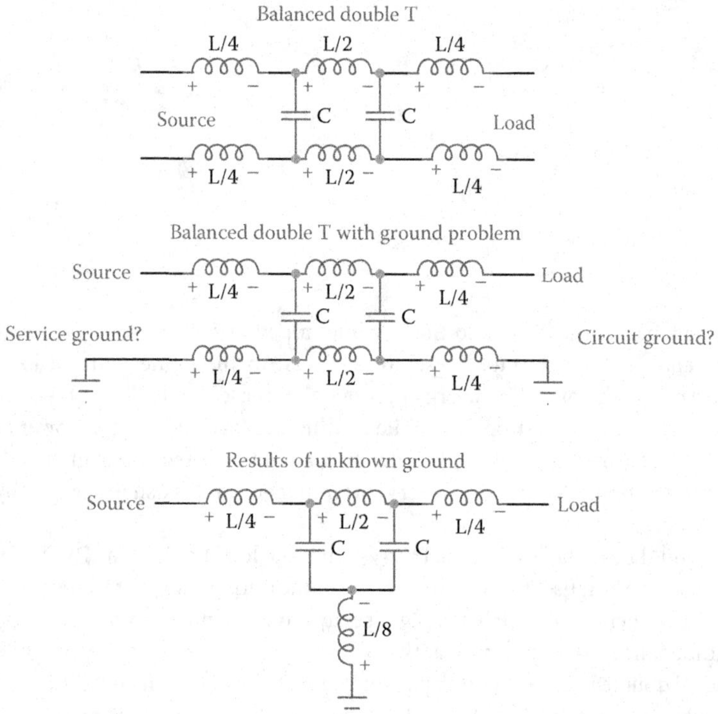

Balanced double T

L/4 L/2 L/4

Source C C Load

L/4 L/2 L/4

Balanced double T with ground problem

Source L/4 L/2 L/4 Load

C C

Service ground? Circuit ground?

L/4 L/2 L/4

Results of unknown ground

Source L/4 L/2 L/4 Load

C C

L/8

FIGURE 7.4 Balanced double T with a ground problem, and results of unknown ground.

Source L L Load

C C

Double L

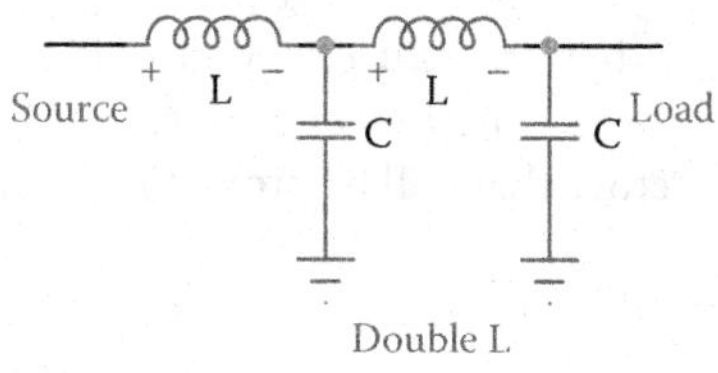

FIGURE 7.5 Double L filter.

This could create a larger peak-to-peak voltage drop feeding the switcher where the peak-to-peak voltage would be at the switcher frequency. The capacitor reactance must be lower than the input impedance of the switcher. The double L can be used as long as the drop is not excessive or the switcher frequency is high enough. This statement is true even though the attenuation, or filter loss, is improved for the DC source. The L and multiple L work well in higher power applications (Figure 7.5). Again, to balance the L filter, split the inductors and put the other half in the neutral. However, the double L has 24 dB of loss per octave. Once again, for a balanced filter topology, it is important to validate the filter and power system ground structure so that there are no hidden grounds that change the circuit configuration.

7.4 The Typical Commercial Filter

This is the type used in test equipment, computers, and other commercial electronic equipment. Here the manufacturer has to pass tests for UL, TUV, VDE, CSA, and the FCC. The tests are conducted by EMI test houses that help the manufacturer with all the documentation needed for the various agencies. These filters are the balanced π type and are often purchased from outside suppliers and are often built offshore.

The filters are mainly common mode in appearance because of the common-mode inductor and two feed-throughs (or these could be 'Y' caps to ground), with a capacitor across the input and output from hot to neutral and two other capacitors to ground. The differential mode is created by the common-mode leakage inductance and the two line-to-line capacitors. These filter types typically do not have a large amount of loss. The feed-through capacitors make the two output terminals and must meet the leakage-current specifications for whatever agency has the toughest requirement. The leakage inductance is often made high by adding washers to the center of a pot core separating the two windings. This is also accomplished by winding the two windings as far apart as possible on a ferrite toroid core. The feed-through capacitors are grounded directly to the case or, if these are Y caps, they must be soldered to ground (Figure 7.6). Some of these techniques add additional differential-mode loss by increasing the leakage inductance or adding inductors to both lines. This then makes a balanced π type with both differential and common mode. This all works because the losses specified for the FCC start at 450 kHz. The inductors and capacitors can be quite small to accomplish these tasks.

A point to consider is that the current through the leakage inductance cannot saturate this inductor because most of this is not through the core, but in the surrounding air. The reason for this is that the common-mode inductor is wound on ferrite or nanocrystalline cores with high A_L values. Some of these filters do not use the feed-through type of capacitor, so the circuit changes to that in Figure 7.7. These capacitors are less expensive, but the self-resonant frequency is lowered by the added lead length.

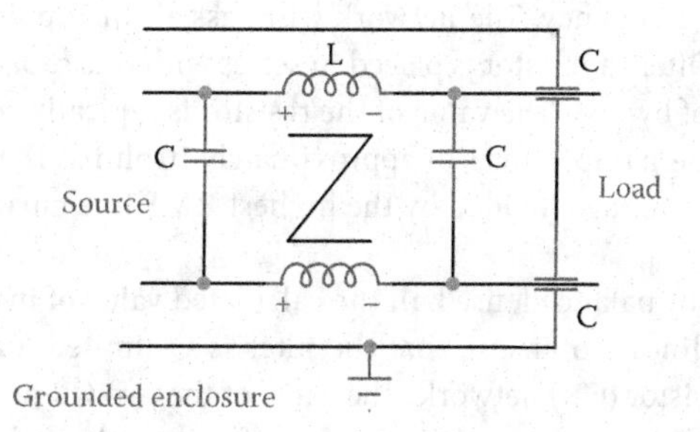

FIGURE 7.6 Typical commercial filter with common mode core and feed-through capacitors.

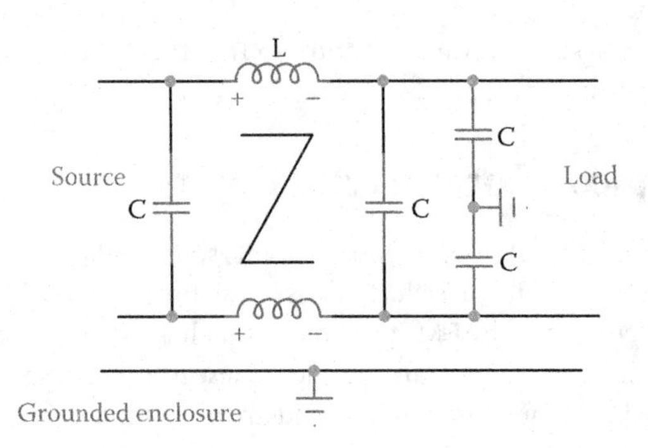

FIGURE 7.7 Commercial filter with Y caps.

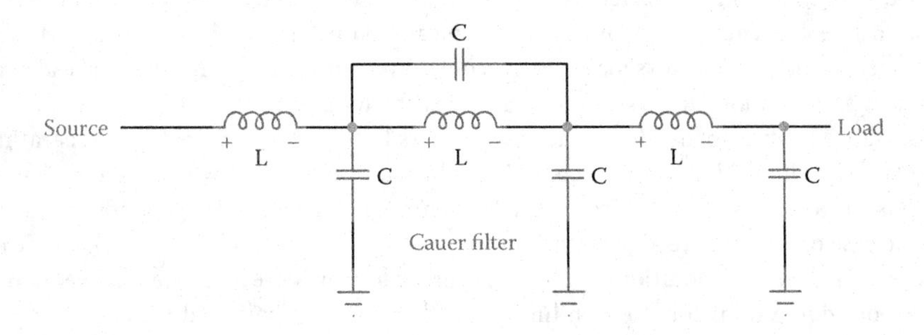

FIGURE 7.8 Cauer filter.

7.5 The Cauer Filter

The Cauer, or elliptic, filter is best used in very-low-impedance circuits (Figure 7.8). These filters are usually used with multiple L and T structures. In any case, a capacitor is normally shunted across one of the central inductors. This is used to reduce the resonant frequency amplitude (Q) such as 14 kHz, for example. The network is tuned to slightly above the problem frequency. Even though the unwanted resonant rise may be somewhat reduced, the center section of the filter structure will not be in the circuit much above this problem frequency. The network will pass all the upper frequencies via the parallel capacitor. Often, a resistor is placed in series with this capacitor, and the resistor limits the amount of bypass. The value of the resistor is typically equivalent to the filter characteristic or design impedance, or approximately 10 ohms. The design impedance is the lowest RMS line voltage divided by the highest RMS line current in MIL-STD-461 specifications.

If this filter is to be balanced, use half the calculated value of inductance on both the power and return lines. To ensure that the filter is Q-limited, parallel each inductor with a capacitor-resistor (CR) network. The value of the two CR networks uses twice the capacitance and half the resistor value for that of an unbalanced configuration. These CR networks would, again, be tied across the two inductors.

7.6 The RC Shunt

Another technique for limiting the filter Q is preferred to the Cauer, but it is better when used in high-impedance, low-current circuits. This filter, called the RC shunt, uses fewer components and is automatically balanced across the line to start with. The RC shunt is shown in Figure 7.9.

The RC shunt is formed with a capacitor and a series resistor. RC shunts are used for two purposes. The first is to provide damping to an LC-resonant circuit such that the resultant circuit $Q < 2$. The resonance effect occurs when inductive and capacitive reactances are equal in absolute value; the Q factor is a measure of the resonance peak amplitude. Therefore, we can say that $Q = j\omega L/R_{DC}$. The second reason to use the RC shunt is to ensure defensive design against higher resonant frequencies due to inter-structure impedance mismatch, parasitic effects, etc. The RC shunt can be applied to a given higher frequency resonance, and the effect of the shunt will be to reduce the peak amplitude of the frequency of interest so that the filter loss is within defined dB limits. It is important to note that adding an RC shunt at the output of a filter will change the filter output impedance, and this must be considered in terms of stability in PWM power supplies that exhibit incremental negative resistance.

Typically, the filter has a resonant rise or pole-Q frequency that is lower than the "trouble" frequency. This is especially true if the filter is a multiple filter such as a double or triple L, π, or T. Usually, the number of resonant rises is one less than the multiple numbers, meaning that the single L, pi, or T would not have any resonant rise, but the quad would have one less, equaling three. This holds true only if the circuit Q is low enough. The higher Q filter has a resonant rise for each network. To implement the RC shunt, as in the case of a trouble frequency, it is necessary to determine the frequency of the lowest resonant rise and pick the capacitor value at this frequency that equals the filter design impedance of the system. This will attenuate each resonant rise above the first and also the trouble frequency. If the lowest resonant rise frequencies are of no concern—well above the fifth harmonic of the power line frequency and well below 10 kHz—choose the capacitor to equal the design impedance at the trouble frequency.

As an example, the design impedance R_d is calculated by dividing the highest current required by the load into the lowest anticipated line voltage. In this case, assume 100 V is the lowest line voltage at 10 A. This is the highest current at this lowest line voltage. The resonant rise frequency is 4 kHz. Then

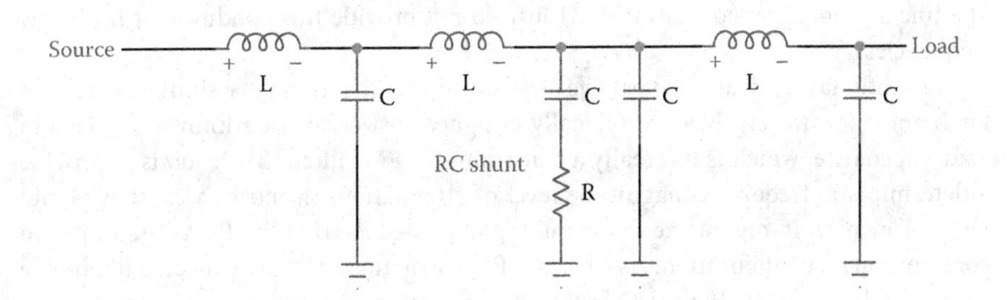

FIGURE 7.9 RC shunt.

$$R_d = \frac{100}{10} = 10 \text{ ohms} \qquad C = \frac{1}{2\pi \times 4000 \times 10} = 3.979 \mu F \tag{7.1}$$

The capacitor is 3.979 μF or 4 μF in series with the 10-ohm resistor tied across the line. This should help to attenuate the resonant rise at 4 kHz and reduce the dB magnitude of any higher problem frequencies (within a certain range). If the resonant rise of the filter is located at a frequency of little concern, it may be beneficial to tune the RC shunt to match any higher problem frequencies. If we were to assume a problem frequency of 14 kHz, for example, we would then calculate the value of the capacitor needed at that frequency.

$$C = \frac{1}{2\pi \times 14{,}000 \times 10} = 1.137 \mu F \tag{7.2}$$

The 10-ohm resistor is in series with a 1.2-μF capacitor tied across the filter. This lends itself to multiple L filters, where the preceding network can be tied across any one of the capacitors. The closer it is to the load, the more it tends to minimize or reduce the impedance swings of the load. As a typical design approach, the implementation of an RC shunt required for filter dQ purposes (the need to control circuit Q and not higher problem frequencies) would normally require that the resistor be equal to the design impedance or the filter characteristic impedance. The shunt capacitor blocks DC current and avoids significant dissipation in the resistor. To allow the resistor to damp the filter, the capacitor should have an impedance magnitude that is sufficiently less than the resistor at the filter resonant, or pole-Q, frequency. This is typically defined at four times the value of the differential-mode capacitor and is discussed in detail in chapter 19.

7.7 The Conventional Filters

EMI filter houses rarely design conventional wave filters such as Butterworth or Chebychev filters, although some applications require the use of these filter types. One major difference when implementing this type of filter is ensuring that both source and load impedance are equal for maximum power transfer. This must be true for the filters to work properly, especially for the low-frequency losses required by the military. The line and load impedances in EMI just do not provide this condition at these low frequencies.

Conventional filters are also passive lossless filters in that they have similar structures for L and C; however, they are typically designed to work with a known source and load impedance, which is not really a constraint for EMI filters. Their job is to provide a deterministic frequency-magnitude level of attenuation. In most cases, they would be implemented in one of the following topologies—LP, HP, BP, BR, AP—and not in common-mode configuration. As a regular filter structure, they are expected to provide loss at the designed –3-dB pole-Q frequency while attenuating higher-order frequencies as a function of the frequency-magnitude slope of the filter.

There are several ways in which to design a regular wave filter. One such method is the coefficient-matching technique, where a given amplitude response such as Butterworth may be defined in transfer function form $H(s)$, with both source and load impedances present. Thereafter, the filter transfer function is matched to the general quadratic expression that describes a second-order system. If we consider Figure 18.8 (chapter 18), the transfer function of an L-type, two-pole, low-pass filter with both source and load is

$$H(s) = \frac{\frac{1}{LC}}{s^2 + \frac{L + CR_L R_S}{LCR_L} s + \frac{R_L + R_S}{LCR_L}} \tag{7.3}$$

From equation (7.3), we can see that this filter has two complex poles.

$$H(s) = \frac{k\omega_n^2}{s^2 + 2\zeta\omega_n s + \omega_n^2} \triangleq \frac{1}{(s - p_1)(s - p_2)} \tag{7.4}$$

The normalized poles of $H(s)$ are located at $p_1, p_2 = -0.707 \pm j0.707$.

From here, we define the relationship between the damping coefficient ζ and the filter Q

$$2\zeta\omega \equiv \left|\frac{\omega}{Q}\right| \Rightarrow Q = \frac{1}{2\zeta} \tag{7.5}$$

In equation (7.3), normalize both the source and load impedances to unity,

$$H(s) = \left. \frac{\frac{1}{LC}}{s^2 + \frac{L + C}{LC} s + \frac{2}{LC}} \right|_{R_L = 1}^{R_S = 1} \tag{7.6}$$

From inspection of equation (7.6)

$$2\zeta\omega = \frac{L + C}{LC} = \sqrt{2} \qquad \frac{2}{LC} = 1 \qquad k = \frac{1}{LC} \tag{7.7}$$

From equation (7.6), we can define scaling terms for $L = C = \sqrt{2}$ and the filter $Q = 0.707$. It is intuitive to see that $\omega^2 = (LC)^{-1} = 1$ and $|\zeta| \equiv |Q| = \sqrt{2}^{-1}$.

From looking at equation (7.3), we can also see that if the source and load impedances were to change such that $R_S \neq R_L$, both the filter Q and passband gain would be modified, and the filter would not provide the required level of performance. In the case of EMI filters, we really don't care about passband attenuation, as we are not dealing with signals that must be subject to discrimination. The coefficient-matching method is also discussed in chapter 18.

8 Filter Components—the Capacitor

The differential-mode components must be of the high-Q type. The individual component Q must be high, while the circuit Q must be lowered to the point where the filter circuit is able to provide effective damping of these parasitic oscillations. Additional circuits are often added to lower the Q below 2. Throughout this book, the term sometimes used for this is *dQ*, which uses either the RC shunt or a series LR network in parallel with the filter inductor. The term *dQ* is pronounced "de Queing." Most EMI houses buy their capacitors from qualified suppliers to ensure robustness, effective SRF, and reliability. EMI filters demand specific types of capacitors for different levels of implementation.

8.1 Capacitor Specifications

The main specification for the capacitors is MIL-STD-15573. The capacitor must meet various voltage ratings and for AC capacitors, the level must be 4.2 times the RMS voltage of the system. For example, in a 220-V RMS system, the capacitor must be designed to handle 924 V, usually rounded up to 1,000 V. For the DC capacitor, the multiplier is 2.5 times the system voltage. For example, a 50-V DC capacitor must be designed to handle 125 V DC, probably made up to 150 V. Also, the RMS peak voltage and the maximum applied DC voltage is used to determine this, not the nominal or average voltage. If this is a 120-V AC system, then we can safely assume ±10%; the peak value of 132 V is multiplied by 4.2 = 554 V, which is the final test voltage for the capacitor. These are tested in the process of manufacturing, and the first test is started after soldering (swedging), possibly at 1,200 V. There may be specifications on certain creepage distances and corona specifications if the voltage is high enough. In the build to print–type of filter specification, the capacitor values may be specified as a minimum or maximum: something like ±10%. Regardless of how well the filter satisfies the insertion loss, the filter will be returned as being out of specification if these values are not adhered to. In most filters, if the capacitor value is reasonably higher, the filter will work much better and give more insertion loss, but that will not suffice if the value is limited. The specification writer does not understand this.

8.2 Capacitor Construction and Self-Resonant Frequency

Capacitors, like everything else, have improved over the years. Better materials and newer techniques have improved the self-resonant frequency (SRF) of polyester (Mylar), ceramic, and other dielectrics for capacitors. Feed-through capacitors using these polyester and Mylar dielectric materials have higher SRF values over what was previously available. The old principle of short lead length still applies. If a long lead is necessary, make the inductor the long lead and tie directly to the capacitor. This is demonstrated in Figure 8.1. This is called veeing the cap.

The advantages of the feed-through type are the low equivalent series resistance (ESR) and equivalent series inductance (ESL) due to short lead lengths. This, in turn, means a much higher SRF for the capacitor. The capacitor manufacturers have graphs with SRF's up to I GHz on smaller capacitor values for all types of capacitors. The problem is that the higher the power line RMS voltage becomes, the bigger the margin must be to eliminate creepage and corona. This requires a bigger capacitor, which increases its cost and the cost of the container. As the line frequency increases, the line harmonic current content also increases and at a higher frequency. This also increases the loss due to *dv/dt effects* through the capacitors, which increases the ESR losses. If the margins were near the lower limits initially, now that the line frequency has increased, the initial margin may have to be increased. In many cases, the capacitor current is above the design value. In other words, the margin size is not a function of frequency unless the capacitor is close to the margin limit.

Some filter companies use the large nonpolarized can-type filter capacitors. Some of these are oil impregnated. These may be very good capacitors for power supplies and other applications requiring nonpolarized capacitors, but watch out for their self-resonant frequencies. Most of these types have a very low SRF, on the order of 50 kHz or less. This is where the old power supply technique—the "ye olde paralleling capacitor trick"— may come in handy. We need to be beware of the resonant rise due to the larger capacitor above the SRF, and therefore being inductive dominant. At higher frequencies, this is effectively an inductor in parallel with the smaller capacitor, and a recipe for

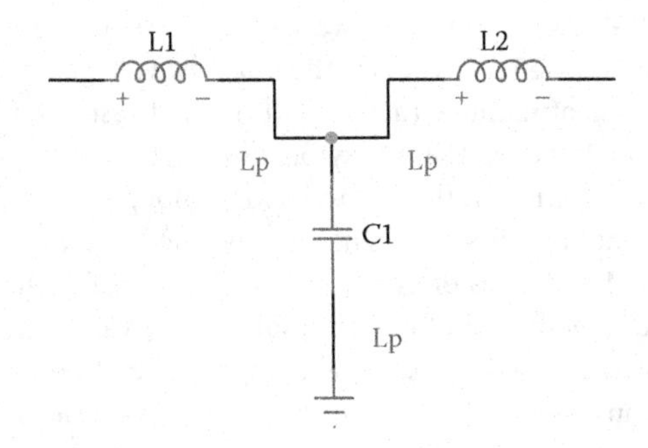

FIGURE 8.1 Veeing the cap.

possible failure. The circuit will behave as a parallel tank circuit with high impedance at the resonant frequency. An article in the IEEE magnetic manuals showed that only 6 dB of gain is realized by these two capacitors in parallel. This theory assumed that the lead length would be close to the same. Consequently, this would almost make the second, smaller capacitor have an ESR and ESL the same in proportion to the original. A feed-through type for the second capacitor would guarantee a workable system because of the very low ESR and ESL and much higher SRF of this high-quality type of feed-through capacitor. Experience shows that paralleling the capacitors often gives serious peaks because of the feed-through capacitor oscillating with the ESL of the original capacitor. The purpose of this nonpolarized style of capacitor is to handle the low-frequency requirement of the EMI filter, and it should not be required over their SRF point. These capacitors often cost less for the capacity and working value; 10, 15, 20, and 30 μF at 480 V AC are available. Another method for lowering the ESL and raising the SRF of the capacitor is as follows.

8.3 Veeing the Capacitor

In Figure 8.1, there are four inductance elements that are often deemed insignificant. This is where the three inductance terms join at the top to represent some tie point or splice, with a fourth at the bottom. The two capacitor lead lengths will add to the ESL and ESR of the capacitor that lowers the capacitor's SRF. Tying the inductor self leads directly to the capacitor improves the capacitor's SRF. The two inductor lead lengths add a small value of inductance increasing the inductance slightly. This slightly increases both inductor values; however, they are orders of magnitude lower in value. This is similar to the concept of keeping the lead lengths as short as possible. These leads are directly connected to the capacitor and are in midair, which lowers any added capacitance to ground. This principle cannot be applied to PCBs due to the fact that all traces are close to the ground plane, which creates capacitance.

Although this technique has been around a long time, the new name for it is "veeing the capacitor." This is the old "keep the leads as short as possible" trick, especially on the capacitor side. The leads on the inductor just add a small amount of additional inductance to the two inductors. However, the leads facing the capacitor, the vertical leads, increase the ESR and ESL of the capacitor, lowering the SRF.

8.4 Margins, Creepage, and Corona—Split Foil for High Voltage

It is important that margin, creepage, and corona are fully understood. The Margin requires a space or a gap to keep the two capacitor plates apart. The space is calculated using 16 volts per 0.001 inch minimum (16 volts per mil). This leaves an area between the dielectrics without a plate - see figure 8.2. If the working voltage is 220 V AC, use 3/32, or 0.093 of an inch. The two Dielectrics and two foils, or plates, are wound on an arbor for the required turns, then the left foils are swedged (soldered) together and so are the right foils swedged together. These become the two plates of the capacitor. If the margins are adequate, everything is okay. Figure 8.2 shows the amount being wound to

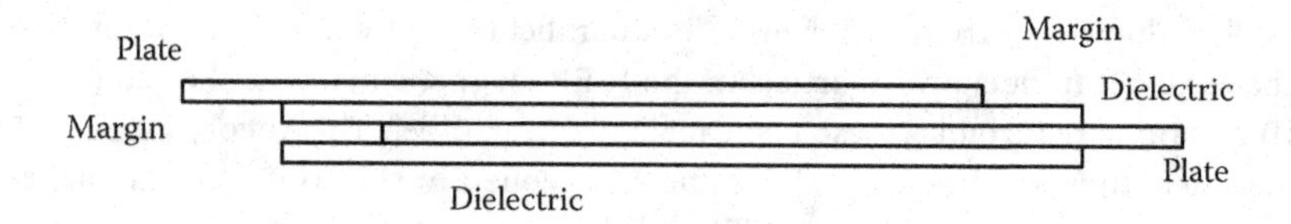

FIGURE 8.2 One layer of winding showing margins.

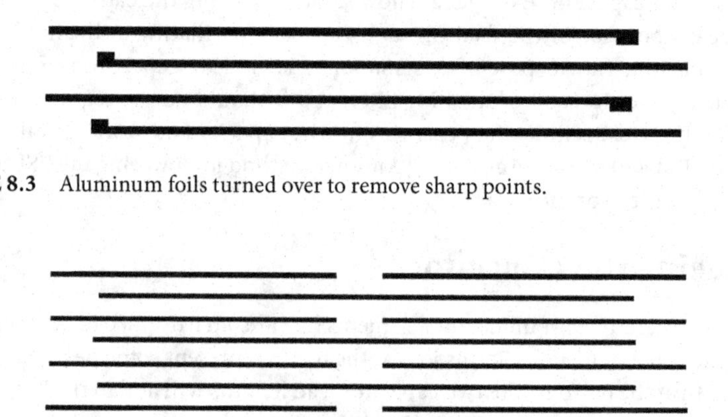

FIGURE 8.3 Aluminum foils turned over to remove sharp points.

FIGURE 8.4 Plated in series for high-voltage application.

build this capacitor. The drawing shows two aluminum plates, two dielectrics, and two margins all being wound at the same time. Actually, there could be three dielectrics at the same time, giving a total required due to the voltage. This means six dielectrics and two aluminum foils—a total of eight layers—all being wound at the same time.

If the capacitor is used at a voltage over its design rating, the voltage will creep over the dielectric slowly, carbonizing a path as it goes until it arcs to the swedged plates, thereby ruining the capacitor. This will take some time to happen, and the carbonized path will grow slowly over time.

Corona, on the other hand, is related to high voltage. To avoid corona, smooth surfaces are required, not sharp points. Capacitor winders require splitters, which spilt dielectric material and the foil to the proper width required for the capacitor. In this way, the capacitor manufacturer buys only the wider widths and split the dielectric material and foil for the desired width. The point here is that if the splitting process forms points in a high-voltage section, corona will develop. One way to prevent this is to fold over the edge on the margin side, forming a rounded smooth end facing the margin. Using this technique, the capacitor becomes thicker with an increased diameter, but this does eliminate all sharp points on the foil (Figure 8.3). In high-voltage capacitors, the plates, or foils, are wound in series to divide the voltage.

In Figure 8.4, the dielectric is not shown along with the margins, and the dielectric must be in the gap betweens the two plates. As seen in Figure 8.4, an extra section divides the voltage between the two main sections to accommodate the higher voltage requirements. However, the active area is reduced. With two sections shown, the voltage is divided by 2, but the capacitor value must be multiplied by 2. The gap in the middle to separate the two foils must be greater than the value used for the margin. Use 32 mils per volt to calculate this distance, and this distance adds. The right and left ends are swedged together. There is no connection to the center foil or plate. These types of capacitors are large, and there may be several in series with more than one split or gap.

8.5 Capacitor Design—Wrap-and-Fill Type

Three methods are used to build these capacitors and two subgroups. The initial method was to attach leads to the foil at the start before wrapping, or winding, the capacitor. These capacitors are now called inductive capacitors or "chicklets" because of the current flow through the foil to charge or discharge the entire length of the capacitor. The ESR and ESL are very high in these chicklets, so the SRF is quite low. As an example, for a foil 0.00023 in. thick and 0.5 in. wide, the cross-sectional area in square inches is 0.000115. Dividing this by 7.854 × 10E-7 (the standard conversion from square inches to circular mils) gives 146.42 circular miles. Divide 17 (the conduction of aluminium foil) by 146.42 gives 0.1161 ohms per foot. If the 0.23-mil aluminum foil is 100 ft long—an average size capacitor—the average length of electron flow is half this, but there are two plates. Thus the resistance is 11.61 ohms. The inductance would be on the order of 45 μH based on approximately 30 meters. It is quite probable that these are not built any longer. Moreover, these should never be used for the main EMI filter capacitors, although they can be used for RC shunts if the frequency is low enough. They can also be used for the Cauer-type shunts for low-frequency tuning capacitors, again depending on the frequency. The tuning frequency must be at least half the chicklet's SRF. If the filter has high-frequency problems where the RC shunt could aid, the quality of the capacitor dictates the higher quality of the extended-foil type (discussed later in this section) instead of a chicklet.

The two subgroups are metallized and foil. For the metallized style, the dielectric is coated with a thin spray of metallized aluminum that becomes the plate of the capacitor. The other subgroup is aluminum foil. The typical aluminum foil thickness is 0.00023 inch, or 0.23 mil. This foil is much thicker than the film, which is measured in microns. Foil will carry much more current and is therefore better for pulse applications and EMI filters where there are high harmonic currents from off-line regulators and similar harmonic sources. Most EMI filters are built with this type of construction and have a thicker aluminum foil if higher currents are expected. Just as the dielectric can have several layers for higher voltages, the foil can either be thicker or also layered. The metallized film has several advantages, however. This capacitor can be much smaller for the same capacitor value, and this type is self-healing. All dielectrics have small pinholes throughout their length. When the applied voltage stresses the film, the film often shorts out through one of the pinholes, causing the film to melt. The aluminum on the metallized film will then re-form, making the capacitor self-healing. Another advantage of this subgroup is that the aluminum can be sprayed on both sides of the metallized dielectric. This adds to the smaller size, promotes self-healing, and is better for extended life.

To avoid the pinhole problem, several dielectrics are used. Typically, the thickness is 0.24 mil each, giving a total dielectric of 0.48 mil. The odds of two holes occurring together are remote. Film capacitors can be used for DC as well as 50- and 60-Hz filters, but they are not recommended for 400-Hz or higher frequencies. However, the capacitor facing the load must be foil, especially for any load that creates pulses such as PWM-based converters.

The type used mostly in the EMI filter is the extended-foil type. The foil extends beyond the winding arbor so that one plate, or foil, extends to the left and the opposite plate extends to the right. The area on the right of the first foil, or the left of the second foil, makes up the margins for both plate ends. It should be obvious why this type is called the extended-foil type. The extension is typically 3/32, and each end is soldered (called swedging) for the contacts. The ESR and ESL of extended-foil capacitors are both very low because the current flow travels only the average width of the capacitor foils, called the gauge, or height of the capacitor. Also, the turns are in parallel, so the average diameter times the number of turns times the thickness gives the square inches of the aluminum. Convert this to circular mils (divide by 7.854×10^{-7}), and divide this into 17 for aluminum (10.374 for copper) to obtain the ohms per foot. Dividing this by 12 inches per foot and multiplying by the gauge (the height in inches) gives the approximate resistance of the capacitor, which is a close approximation of the ESR. Inductance is approximately 1.5 μH per meter. Divide 1.5 by 39.36 inches per meter to get 38.1 nH per inch. This term multiplied by the gauge gives a first-order approximation of the ESL. It should be apparent why a large capacitor diameter to length ratio, or gauge, is desirable. A 2:1 ratio of the diameter to the height, or gauge, is the optimum ratio. If the ratio increases too much over this optimum, the capacitor starts to wobble on the winding arbor, and the active plate area is diminished. The capacitance value drops, but the gauge is also more than desired. The SRF is very high for this optimum type and should be used by the EMI filter designer.

The last method—called the tab type—is similar to the extended-foil type but is used for applications with high capacitor current. The size and number of the tabs depend on the current. These tabs are thin strips of conductor that are placed in the winding as the capacitor is being wound. These tabs are inserted—one for each plate—every so many turns, and a sufficient number of degrees is added to each tab so that the tabs end up uniformly spaced around the sides of the capacitor. The tabs extending out of each side of the capacitor are then folded over and soldered together to form the contact.

Most new people in the EMI design arena are knowledgeable about power supply design, where the main concerns about the capacitors are their working voltage and capacitor value. Others are familiar with derating of the capacitors, and this is certainly helpful. The point is that AC capacitors must be designed to handle the total AC currents at the line frequency and the odd harmonics. If this is a single L or T, the capacitor should be of the foil type, not metallized film. In multiple filters—double or triple of any topology—at least the last capacitor, or the capacitor closest to the load side, should be of the foil type.

Some capacitors are not designed properly to handle the full AC current flow or were designed for DC applications. This is especially true if the capacitors are the metallized film type. The capacitor designer may not have designed the capacitor to handle the total harmonic current. Capacitors must handle the harmonic current from either the line frequency side or the load side. This is especially true for the harmonic current created by the off-line regulator or any power supply using a capacitor input filter. The foil making up the capacitor plates can be too thin to handle this current. This raises the ESR losses, so the capacitor will heat and fail in the months ahead. There is an issue here. Companies will often take the failed unit apart and replace the blown capacitors with the same capacitor, stating that they are not aware that the capacitor has the wrong rating for this application. Commonly, the capacitor was designed for DC operation and not for AC by the capacitor manufacturer, and was selected in error by the original filter designer. The filter capacitor must be selected to handle the harmonic currents of the off-line regulator and any other pulse type with high harmonic currents.

9 Filter Components—the Inductor

The EMI filter requires what is termed a *soft core*. This really means that the core is driven into saturation slowly rather than abruptly, as required for pulse transformers and magnetic amplifiers. A hard core can be made soft by gapping the core. This technique tilts over the *BH* curve to the right in the first quadrant (Figure 9.1), making the core a soft core—harder to drive into saturation.

9.1 Inductor Styles and Specifications

Inductors for EMI filters come in several styles: tape wound, toroids, C cores, and slugs. E cores, pot cores, and RM cores are rarely seen. There are several subgroups. For toroids, the styles are ferrite, powdered iron, MPP, high-flux (HF), and Kool Mu (the old Sendust material). For C cores, or cut cores, the styles are various steel mil thicknesses as well as steel types. The same is true for the tape-wound toroid. For the slug, other than size, there are various mixes. As far as specifications are concerned, some filter customers specify the core, the wire size, and the inductor value, usually as a range such as ±10%. Others specify creepage distance or list a specification that calls out the distance. In this case, the clearance listed is between the top and bottom wires to the core. For the tape-wound toroid this does not apply. This is more like a margin along the coil form. Sometimes a wire size is specified.

9.2 Core Types

9.2.1 Power Cores

The core types used for EMI filters are often molybdenum permalloy toroids, sometimes the HF and Kool Mu types. The main advantage is that all three types of cores can have very high Q values. MPP cores are used throughout EMI filters. These cores saturate at 7,000 gauss and are usable to 3,500 gauss. Their primary purpose is for the differential-mode filter inductors. They do not require gapping because they have distributed gaps throughout the core. These are easy to design using the given A_L

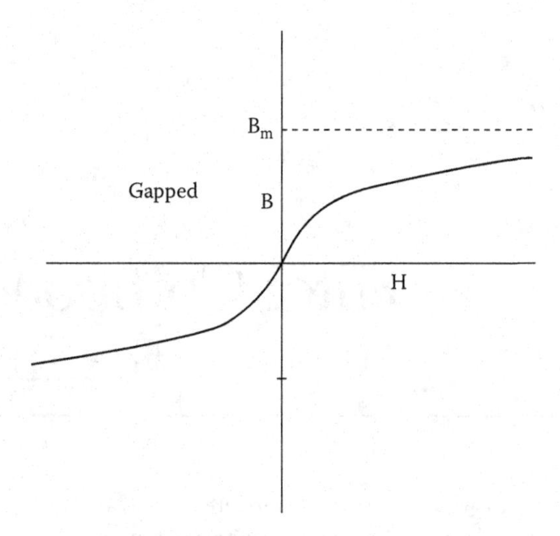

FIGURE 9.1 BH curve for a gapped core.

values. It is recommended that permeabilities above 125 should not be used for 400 Hz and above.

HF cores are often used in EMI filters; these particular core types saturate at 15,000 gauss and are usable to 7,000 gauss. Their primary purpose is for the differential-mode filter inductors requiring higher currents. Again, these cores do not require gapping because they have distributed gaps throughout the core. It is recommended that permeabilities above 125 should not be used for 400 Hz and above. Also, for DC filters, the flux density should be restricted to 3,500 gauss.

Kool Mu cores are the old Sendust cores using aluminum, and these are also used in EMI filter applications. These cores saturate at 10,000 gauss and are usable to about 5,000 gauss. Their primary purpose is for the differential-mode filter inductors requiring lighter and less expensive toroids. They also do not require gapping because they have distributed gaps throughout the core. It is recommended that permeabilities above 125 should not be used for 400 Hz and above. Also, for DC filters, the flux density should be restricted to 3,500 gauss.

Powdered-iron toroids are also sometimes used, but they often cause trouble at 400 Hz because they can overheat under certain conditions. Powdered iron is the least expensive style of toroids and has a relatively low permeability. The main disadvantage is that the inductance reading is more a function of the bridge drive level than the other toroids. Therefore, the value supplied by the LCR bridge will be different on another bridge because of the differences in drive levels. In other words, this is the least stable core due to the fact that the permeability varies with the impressed current. Regardless, these are often found in EMI filters, especially for low-current filters.

For the cores discussed in this section, there are two main suppliers, Magnetics., Inc. and Arnold Magnetics Technologies.

9.2.2 Ferrite Cores

Ferrite toroids are used in EMI filters primarily for common-mode cores. They should not be used for inductors because they saturate too easily. These cores are driven into saturation at a low H value or a low current. The only exception is when very low current is called for. To be used for inductors, they would need to be gapped, which is not really practicable. Used as an inductor, the filter may pass the specification because the driving current from the tracking generator would be very low. In other words, with light load during testing, these inductors would pass. However, with a working current of 10 A, it is likely they would fail due to saturation. This core has very high A_L values to give high inductance for common-mode loss. The reason is that the line current can be high, and will not saturate the core because the flux cancels in a common-mode core. The common-mode driving current, or EMI, is not strong enough to saturate the core.

9.2.3 Tape-Wound Toroids

Tape-wound toroids can be used for EMI filters, but they are primarily used for transformers. Their permeability is quite high, with a usable flux range of 15,500 to 16,000 gauss. The main disadvantage is their size and weight. (See the section on transformers in chapter 11.) These transformers give low-frequency common- and normal-mode loss along with additional skin-effect losses at the higher frequencies, starting with the fifth harmonic of the line frequency. For use as an inductor within an EMI application, the core must be cut-gapped (the energy is stored in the gap). The inductance equation changes due to the gap. The permeability can divide the magnetic path length, which equals the magnetic resistance (reluctance), but the gap adds to the reluctance, so the total reluctance must be used. The original equation for powder cores is given in equation (9.1). The powder cores have distributed flux density as given by

$$L = \frac{0.4\pi\mu N^2 A_C 10^{-2}}{M_{PL}} \mu H \tag{9.1}$$

When the core is gapped, the magnetic resistance (REL) is increased. The original REL is equal to M_{PL}/μ, which removes μ from the numerator. Now the gap REL must be added to the denominator divided by its permeability, as in equation (9.2)

$$\frac{g}{\mu_{Air}} + \frac{M_{PL}}{\mu_{Steel}} \tag{9.2}$$

The permeability of air is 1, and the permeability of steel may be 2,000. The total REL is essentially the gap (g) in centimeters. So the equation for inductance for a core with a cut gap is

$$L = \frac{0.4\pi N^2 A_C 10^{-2}}{g} \mu H \tag{9.3}$$

The gap (g) length is in centimeters. The gap tilts the *BH* curve over, making the permeability a little more constant or flatter, i.e., less S shaped (see Figure 9.1). The disadvantage is that the Q is low, making the unit a little less efficient. However, these can be used for very high currents, as much as 300 A or more, and the inductance can be 400 μH.

9.2.4 C-Core Inductors

C cores are typically used within EMI filters for higher currents, as these are more than the powder cores are able to handle. These can be designed using 12,000 gauss, while the peak flux density is about 18,000 gauss. Gapping these cores is relatively simple (they are in two pieces to start with), and the inductor does require this because it does not have a distributed gap. Unless the ends are polished, the minimum gap is 1 mil. In other words, the no-gap reading (core against core) results in a gap equal to 1 mil. The inductances of these cores vary from batch to batch. These can be wound using a single wind on one leg or one winding on each leg. The two windings can be made to be series windings or parallel, with half the current in each leg. The latter is wound oppositely on each leg called Left to Right on one coil form and Right to Left on the other. The real disadvantage of the C core is that it has very low Q. Use equation (9.3) for designing C-core inductors. The C core may be gapped using shims of half the gap value in each leg, showing that this would be good for DC operations and would make the core softer. They are straightforward to wind on a bobbin or coil form, whereas toroids are much slower to wind. If many of these inductors are needed, several C-core bobbins or coil forms may be wound at the same time, depending on wire size, whereas a single toroid must be wound one at a time.

9.2.5 Slug Type

The slug is often powdered iron. Typically, this core is an inch in diameter and 1 to 2 in. long (Figure 9.2). These are not as quiet as either the toroid or C core due to magnetic field coupling between each end through air. This is their strong disadvantage. They must be mounted in midair so as not to operate as an inductive heater. Any close-proximity magnetic material, such as the filter enclosure, will become part of the flux path. For example, if it is mounted next to or close to the case, the flux path will be via this case, both heating the case and increasing the inductance of the part due to the low impedance of the new flux path. Slugs have been designed using this type of core with inductance values of 150 μH from Micrometals Inc. These were mounted in midair, hanging by their leads. Then they were potted in place. The effective permeability of Micrometals powdered iron P6464-140 is 5. The equations for this style of inductors are

$$L = \frac{0.8U_{EFF}(RN)^2}{(6R+9T+10B)} \rightarrow N = \frac{1}{R}\left[\frac{L(6R+9T+10B)}{0.8U_{EFF}}\right]^{(0.5)} \tag{9.4}$$

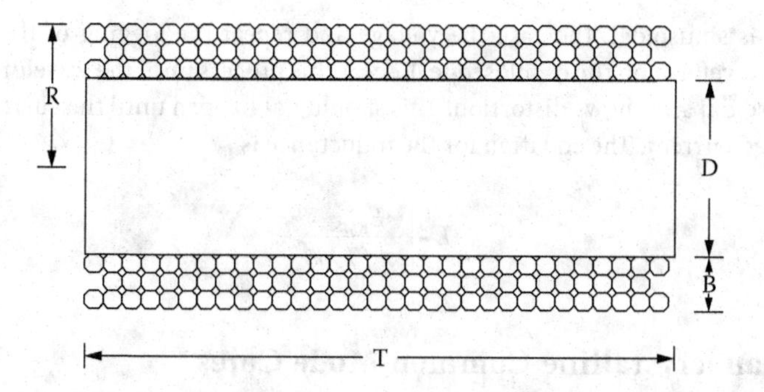

FIGURE 9.2 Multilayered slug inductor.

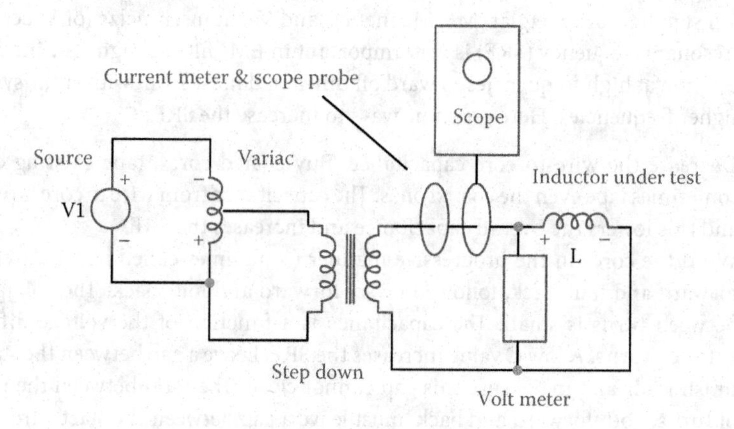

FIGURE 9.3 Measurement setup for the spreadsheet.

All dimensions are in inches, and L is in microhenrys

$$L = \frac{0.8U_{EFF}(RN)^2}{(9R+10T)} \rightarrow N = \frac{1}{R}\left[\frac{L(9R+10T)}{0.8U_{EFF}}\right]^{(0.5)} \quad (9.5)$$

Equation (9.5) is used for the single-layer inductor. If the length T to radius R is above 0.8 ($T > 0.8R$), the accuracy is within 1.0%.

High-current-tape wound or C-core inductors cannot be properly measured by common LCR bridges because the bridge voltage drive is often not high enough. The best way is to use a variable transformer (Variac) from the 50-, 60-, or 400-Hz line (Figure 9.3). For higher current, a step-down transformer feeds the inductor to increase the inductor current. Use a meter to read the current and a voltmeter across the inductor to read the voltage. If at all possible, use a current probe connected to an oscilloscope. Apply voltage and read both the current and voltage; then check the scope to see if the current waveform is sinusoidal. Increase the voltage and repeat the logging of the data, and check the waveform on the oscilloscope. Repeat this process until the waveform changes from sinusoidal and shows distortion. This should not happen until the current is above the required current. The equation for the inductance is

$$L = \frac{E_{RMS}}{2\pi f I_{RMS}} \tag{9.6}$$

9.2.6 Nanocrystalline Common-Mode Cores

These were very expensive in days gone by, but have since dropped considerably in price. These cores have high permeability (μ); therefore, they require much fewer turns and are both lighter and smaller. Typical μ values of 20,000, 30,000, and 80,000 are available. The main suppliers are Metglas, MK Magnetics, and Vacuumschmelze (or Vaccorp).

Self-resonant frequency (SRF) is very important in EMI filter design. The inductance must function at high frequencies to ward off unwanted noise from within the system at these higher frequencies. Here are some ways to increase the SRF.

1. Decrease the wire-to-core capacitance. Buy coated cores, tape existing cores—sometimes tape even the coated ones. The capacitance from wire to core is reduced, and this lowers the overall capacitance and increases the SRF.
2. Wind the cores in the progressive manner (sometimes called pilgrim step)—six forward and four back, followed by six forward and four back. The voltage ratio between turns is small. The capacitance is a function of the voltage difference between turns. A low C value increases the SRF. Leave a gap between the start and finish leads and make sure this gap cannot close. The ratio between the number of turns, both forward and back, must leave a gap between the start wire and the finish wire. This may take some time to get the front-to-back ratio. The disadvantage is that this method can result in an ugly appearance. It is usual to cover it with tape to help hide this feature, as uninformed customers might question the appearance of the inductor.
3. Combining both 1 and 2 above also reduces the capacitance, raising the SRF.
4. Section wind the toroid, such as 15 turns in a clump, then move over a space and do the same: a lot of turns in one area with a space left blank and then followed by the same number of turns with a number of clumps around the core. This makes a strange-looking inductor. The capacitance in each section is small, and there is space between the sections so that the total capacitance overall is reduced. A lower C value again gives a higher SRF value. Tape all the gaps so the various clumps can't slip together.
5. Wind small inductors and connect them in series. The SRF of each core will be higher; therefore the overall SRF will also be higher. These inductors can be placed in quadrature so the magnetic coupling will be reduced.
6. Another way to raise SRF is to use heavier insulation. This forces higher spacing from turn to turn and helps raise the copper farther off the core and decrease the turn-to-turn capacity.

9.3 High-Current Inductors

High currents demand large wire diameters. Welding cable is often employed because it has many strands, such as 833 strands of No. 30. This makes the wire flexible, allowing it to be wrapped around the coil form. Many engineers want to use Litz wire and high-frequency cores, which would increase the cost several orders of magnitude. However, for EMI filters, either Litz or welding cable should be used only for ease of winding the coil form. If the coil is to be tuned, the Q must be high at the tuned frequency. This calls for a core with thinner material and stranded wire so the AC resistance will be low at, and somewhat above, the tuned frequency. For example, a 60-Hz inductor is to be tuned to 25 kHz. Change the core from a National-Arnold Magnetics CA to a CH. This moves the material from a 12-mil to a 4-mil core. The radius of conduction in centimeters is 6.62 divided by the square root of the frequency. Up the frequency to 30,000 Hz and take the square root of it, this gives 173.2. Divide this into 6.62 which will give the radius of conduction in Centimeters. Divide this by 2.54 (centimeters per inch) gives the radius in inches = 0.015. Then 0.015 times 2 gives the wire diameter in inches, thus 0.030. We select AWG No. 20 wire, which is 0.032 and is close enough to 104.2 circular mils. If this inductor must handle 10 A, using 500 circular amps per amp, approximately 5,000 circular mils is needed, which would require 48 strands of No. 20 wire. It should be apparent that Litz may be the best choice. The disadvantage, other than cost, is that fewer turns of this wire can be wound in the available space where 48 individual strands could fit. However, it is quite time consuming to wind this inductor with this many strands.

9.4 Inductor Design

The design of any inductor is often a challenge in many respects. There are techniques that solve for wire size, turns, watts lost, efficiency, and temperature rise. For the tape-wound type, the gap required is given. However, most manufacturers of magnetic cores list the core as giving A_L values in units of millihenrys per 1,000 turns. Others list the values in units of nanohenrys per turn, but rarely for the type of filters designed in this book. This technique can be used for all inductors, including common mode, as long as the A_L value is listed by manufacturer or known by some other means. This was developed from the following for powder cores:

$$L = \frac{0.4\pi \mu N^2 A_C \times 10^{-8}}{M_{PL}} \tag{9.7}$$

In this equation, the required inductance L is known, along with A_C (cross-sectional area), M_{PL} (magnetic path length), and the permeability for the core (μ) that the designer would like to use. What is not known is the term N. These known core quantities make up A_1.

$$A_1 = \frac{(0.4\pi \mu A_C)10^{-8}}{M_{PL}} \tag{9.8}$$

The manufacturer provides the A_l values.

$$L = A_l N^2 \tag{9.9}$$

The required inductance is already known. The A_L value is given, so

$$N_2 = N_1 \sqrt{\frac{L_1}{A_L}} \tag{9.10}$$

where N_2 is the required turns, N_1 is the known turns, and L_1 is the required inductance in the same units as A_L (millihenrys).

For example, the Magnetics Inc. 55351 MPP core is listed as having an A_L value of 51 mH per 1,000 turns. The designer needs 0.8 mH (800 μH). L_1 is equal to 0.8, and A_L is equal to 51. Therefore, 0.8 divided by 51 is equal to 0.01569, and the square root of this is 0.1252. This value times 1,000 equals 125 turns.

$$N_2 = 1000\sqrt{\frac{0.8}{51}} = 125.245 \tag{9.11}$$

The current is multiplied by the circular mils to determine whether the window area is adequate, that is to say, whether the calculated turns will fit within the core window area. This is a small core, but we need 1 A, and using 600 circular mils per ampere would lead to 23 AWG at 650 circular mils per ampere. The 650 times 125, the number of turns just calculated, equals 81,250 circular mils, but the winding factor for this toroid-type core is only 0.4. The 81,250 divided by 0.4 yields 203,125 circular mils.

$$\frac{650 \times 125}{0.4} = 203{,}125 \tag{9.12}$$

Magnetics Inc. lists the window area as 293,800 circular mils, so the windings will fit. The wire size could be changed to 22 AWG, giving 253,125 and still allowing it to fit. This would reduce the copper losses and lower the temperature rise.

9.5 Converting from Unbalanced to Balanced

It is better to balance the EMI filter if the supply or equipment has not already been grounded. Converting from an unbalanced circuit to a balanced circuit is necessary because the equations that solve these circuits are all based upon unbalanced component values.

To balance the filter, as in the 800-μH requirement discussed in section 9.4, use a double L structure with a capacitor, shown in Figure 9.4 as C1. Note that these capacitors in a balanced configuration are no longer from line to ground. In the balanced side, they are connected line to line. The unbalanced drawing to the left would have two L inductors equal to 800 μF and two C1 Y capacitors wired to ground. As a point of reference, the capacitance to ground would be out of limits. The balanced structure has four inductors each valued at L/2. Here the original is 800 μH and now would be half the original value, or 400 μH each. The Y capacitor is the same value, C1, but now it is an X capacitor removed from ground. The SRF value of these inductors is now about $\sqrt{2}$ higher in value. The capacitors are no longer grounded, meaning that they don't impact the to-ground capacitor limit.

Now a common-mode inductor can be added because the circuit is balanced (usually to the left or source side). Four Y capacitors, or feed-throughs, can be added for a common-mode π filter. To meet the specification, the feed-throughs here would be a maximum of 0.025 μF each. Because this is a balanced configuration, the feed-throughs add, since the two Y capacitors (or as in Figure 9.5 feed-throughs) are in parallel at both ends for common-mode loss. Thus, the four terms in L also add inductance to the common-mode inductance. The front two add a half value of 200 μH, and the two load-side inductors add another 200 μH, making a total of 400 μH to aid the common-mode inductor. Note in Figure 9.5 that the four feed-throughs provide input and output terminals for the enclosure, whereas Y caps would not.

The electrical connections are now wired to the four feed-throughs. This is a lot better if there is no "cheat" circuit to ground in the instrument, which would bypass the bottom inductors. Not only do the inductors get bypassed, but now the central X cap has an inductor in series with it, which eliminates any possible filter loss.

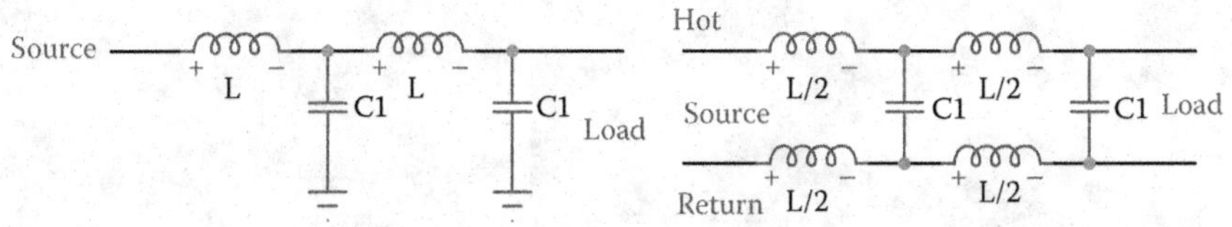

FIGURE 9.4 Convert from unbalanced to balanced.

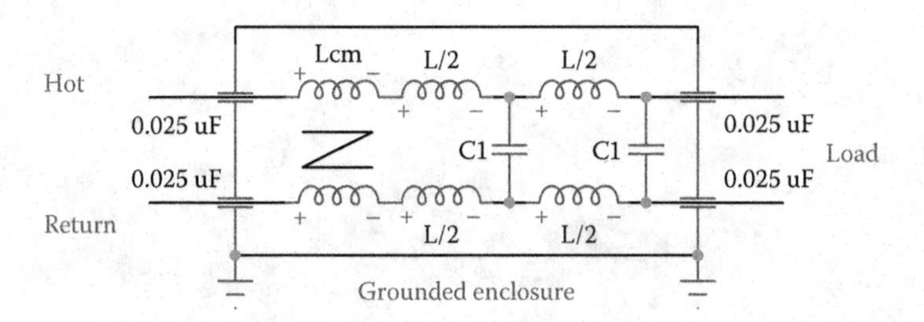

FIGURE 9.5 Balanced EMI filter.

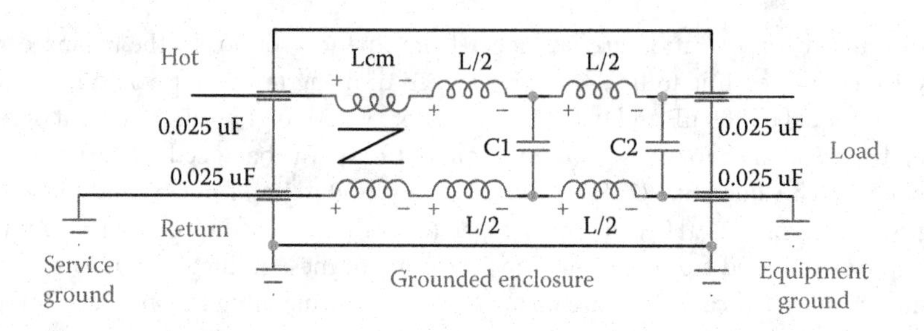

FIGURE 9.6 Compromised balanced filter.

The bottom of C2 in Figure 9.6 (shown as C1 in Figure 9.5) is at ground, which compromises the capacitance-to-ground value. C1 has one 400-μH inductor to ground through the equipment, which is in parallel with the other 400 μH in series with the bottom half of the common mode. This is not a good situation at all. Also, the current through the common-mode inductor is obviously compromised, which will increase the wire temperature rise. The indication that this may have occurred is if the EMI filter readings are way out of specification when attached to the unit requiring the filter. The service ground is normal. It is the unknown ground in the unit that is causing the problem.

10 Common-Mode Components

The causes of spikes or electromagnetic pulses (EMPs) on a power line or conductor are lightning, large inductive equipment shutting off, and magnetic pulses created by nuclear activity. Lightning and nuclear activity create common-mode pulses between power lines or between conductors and ground, whereas the equipment-type disturbances create a differential-mode type of pulse between the lines. On the load side, PWM switching and switching diodes are the leading culprits for generating noise. Any current pulse seen between the load side of the off-line regulator and ground appears as common mode to the line. Adding a transformer or keeping the power supply isolated reduces these common-mode-conducted emission pulses. The one exception is the primary-to-secondary capacitance of the transformer, but this is so small that it can be neglected. Power supply filter capacitors pumping current to ground should be eliminated. To eliminate the common mode, the EMI filter employs common-mode inductors and feed-through capacitors to ground. In some specifications, the capacitance to ground is limited in order to limit the leakage current, and this will reduce the common-mode performance of the filter. The leakage current is the capacitive reactance current flow through the capacitor between line and ground, otherwise known as reactive current.

10.1 Capacitor to Ground

At 400 Hz, the limit of the capacitor to ground is 0.02 μF for MIL-STD-461. Worse yet, the leakage-current specifications for medical devices are often harder to meet. If the device touches a patient, the total system leakage is limited to 100 μA. This means that most of the power supply people want the filter restricted to 20 to 40 μA. It is difficult to have the common-mode losses meet the common-mode loss specifications with capacitors to ground this small. A transformer would help (see chapter 11). There are two schools of thought on this because the relevant specifications are, as usual, not clearly written. Firstly, we could assume that this is the total capacitor to ground. In a three-phase four-wire circuit, the capacitor limit value for 400 Hz is 0.02 μF. This capacitor value would then be shared, or 0.02 μF/4 = 0.005 μF. This means 0.005 μF for each of the four legs to ground.

Secondly, we may assume that this is the maximum per line irrespective of the number of lines. If the system is well balanced, the current on each leg would nearly cancel through these capacitors at the ground point. (See section 10.2 on virtual ground.) It seems that a better solution is with the latter because it makes the job as a filter designer easier. However, ground fault equipment would not allow any capacitance to ground that could produce a current above the current threshold of the ground fault device, or it would have to be in the circuit after the filter. Again, a transformer would circumvent this.

10.2 Virtual Ground

In a two-phase system, where the two lines are 180 degrees out of phase, a virtual ground can be employed. This is the common method for power. Either leg connected to the common ground gives 120 V RMS, and the two outer legs give 240 V line to line, which is really two phase. Assuming that the two line voltages are equal and the two capacitors are equal, the current through each capacitor would be equal. This implies that there would be no current to ground. Whatever current flows in one capacitor to ground, the other capacitor has the same current flowing from ground. The two currents almost cancel. The small ground current is due to the slight differences in the two voltages and the differences in the capacitor values.

The same is true for three-phase systems. If the three RMS voltages are nearly equal and the capacitors to ground are nearly equal, the current to ground will be small. If the line voltages are the same and the capacitors are the same, the ground current is zero. One way to help remove some of the difference currents is to employ the virtual ground technique. This is achieved by tying the junctions of the capacitors together to form a virtual ground. Under ideal situations, the junction voltage to ground would be zero. Tie a capacitor of equal value from the junction to ground. Ground current will flow through the added capacitor based on the junction voltage. This technique further reduces the current on the ground lead. A question arises: Why are capacitors to ground necessary? Common-mode reduction requires them even with a transformer. The three capacitors in two-phase systems and the four in three-phase systems have a reasonably high impedance to ground at 50, 60, and 400 Hz, but what about at 14 kHz and above? The two, or three, capacitors connected between the lines and the junction are in parallel, and are in series with the capacitor from the junction to ground.

10.3 Z for Zorro

Ferrite toroid cores are often used for common-mode inductors because they have a very high A_L value required for these common-mode filters. For common-mode testing, all lines are tied together in parallel and all the differential-mode inductors are in parallel in the balanced design. Capacitors connected line to line are of no value for common mode, but the capacitors from line to ground add in parallel. In the preceding three-phase, four-wire case, the total capacitor to ground would be $0.02\ \mu F \times 4 = 0.08\ \mu F$. To ensure sufficient common-mode loss, a common-mode inductor would be added. The total inductance and the total capacitance to ground normally do not give the required loss. This problem is often solved by placing a well-grounded barrier or shield across the filter center (Figure 10.1). Split the value of the feed-through capacitor limit to ground

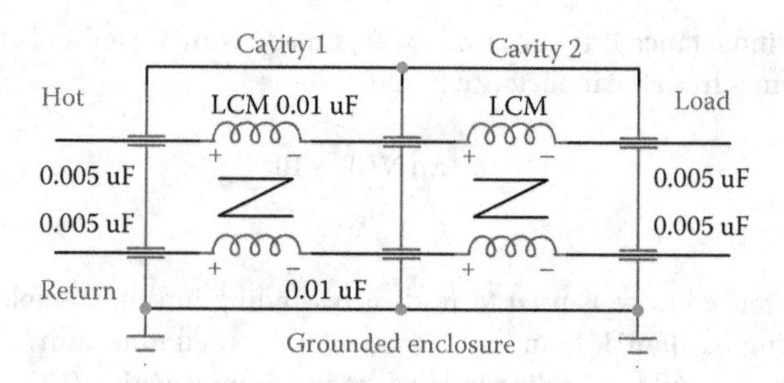

FIGURE 10.1 Double Zorro for common mode.

by four. Two of these smaller feed-through capacitors are then installed in the input and two on the output side of the enclosure. Two more at twice the value are placed in the central shield. For best results with this method, put the Zorro inductor at the low-impedance end following the two front end feed-through capacitors on the line side in cavity 1. The next Zorro is connected to the two central feed-through capacitors in the shield. Try to use an even number of differential-mode filters so that they can be split evenly in the two cavities. Say two L filters are required so that the first cavity would start with the Zorro, followed by the first L, and then followed by the central feed-through capacitors. The second section would be the same. The second L would be located at the output feed-through capacitors.

Of course, the line-to-line capacitors and the differential-mode inductances are not shown because they contribute nothing to common mode. However, they would follow the common-mode inductors as shown in Figure 10.1. It is an option to place a second common-mode inductor in the filter to reduce the overall size of a single core. This approach would place a common-mode inductor in the front cavity and another in the second cavity. This technique forms a double π filter in the common-mode filter section, thereby reducing the common-mode inductor greatly in size. This will ensure that the common-mode loss specification has a much better chance of being met. Remember that the inductor's magnetic fields buck, or cancel, for the differential mode, and they have a high magnetic gain for the common mode.

For the three-phase Y filters, another set of windings is required, thereby demanding a larger core for the same form. These are excellent for creepage; however, the leakage inductance is greater, which increases the differential-mode inductance. These separate windings create approximately 0.5% to 1% leakage or differential-mode inductance. There are cases when this leakage inductance is increased so that the filter has increased differential-mode loss. In most cases, it is maintained at 1% by design. Some designs have transitioned from the separate core windings to the quadfilar type to eliminate ringing that may be caused by leakage inductance and the stray capacitance within the inductor and other wiring.

10.4 Common-Mode Inductor

The common-mode inductor has one value assigned to it. The inductance value is written above the Z for Zorro, say 10 mH, which is a typical value. Either winding should read the indicated inductance if the measurement is made with a good inductance bridge. The reason comes from the inductance formula

$$L = \frac{0.4\pi\mu N^2 A_C \times 10^{-8}}{M_{PL}} \tag{10.1}$$

If an inductance bridge is used to read each winding in this example, the reading of both windings is 10 mH. If an inductance bridge is used to measure the two windings' aiding, what would the aiding inductance be? From equation (10.1), the turns are squared, so twice the turns give four times the inductance, or 40 mH. However, the windings are split, half on each line, so each half is 20 mH. These two windings are actually in parallel, so we are back to 10 mH. In other words, if either winding would measure X and if both are measured in parallel, aiding would measure $4X$. If this was measured with the windings opposing, it would read the leakage inductance. Another way to look at this is to think that the windings are bifilar. Each gets half the current, one wind or the other, plus the two windings in parallel all give the same reading, here 10 mH.

A common-mode inductor using a ferrite toroid core can be designed using the A_L value of the core. This would be used, though only for bifilar types requiring very low leakage inductance. The only difference between designing the differential mode and the common mode is that the winding window fill factor is no longer 0.4 but now is reduced to 0.2 to fit the two windings. Divide the number of leads into 0.4 to get the winding factor.

The single-layer ferrite toroid winding can also be found when the core I_d and the wire size (American wire gauge, AWG) needed to handle the current are known. If the wire diameter, W_d is specified in the same units as the I_d of the core, then the total turns, N_t is equal to

$$N_t = \frac{\pi(I_d - W_d)}{W_d} \tag{10.2}$$

If the wire diameter is much smaller than the core diameter, this approaches

$$N_t = \pi \frac{I_d}{W_d} \tag{10.3}$$

Divide the turns by two, or whatever the number of wires is, and use the lower integer value to solve for the inductance. Knowing the A_L value of the core and the number of turns that the core can support, the inductance value can then be calculated. This would be in a single layer wound less than halfway around. If this is not greater than the needed inductance, pick another core, usually the next size up, with a larger I_d. Once a core is found for which the inductance is somewhat larger than needed, resolve the number of turns required by using the normal A_L equation. Find the number of turns necessary and use the higher integer. For example, 8 mH is needed with a current of 2 A peak. The

AWG is No. 18 picked for the current specified. The diameter of the wire in inches is 0.0429. Now pick the core. Here, 42915-TC is selected from Magnetics Inc. The I_d of this core in inches (same units) is 1.142

$$N_t = \pi \frac{1.142}{0.0429} = 83.629 \tag{10.4}$$

Divide this figure (83.629) by 2, obtaining 41.81, and round off to the nearest integer. Here the integer is 42 turns. The A_L value of this core is 3,868 mH per 1,000 turns using F material (one of magnetics Inc.'s ferrite materials). Here, the backward formula for A_L is used for 42 turns.

$$L = \frac{A_L 42^2}{1000^2} = 6.502 mH \tag{10.5}$$

This is a little low, so a larger core or a material with a higher A_L must be selected.

The diameter of the wire is still 0.0429 in. Pick a new larger core such as the 43615-TC from Magnetics Inc. The I_d of this core, in inches (same units), is 1.417.

$$N_t = \frac{\pi \times 1.417}{0.0429} = 103.768 \tag{10.6}$$

Again, divide this by 2, obtaining 51.88 turns, and use the lower integer.

$$L = \frac{4040 \times 51^2}{1000^2} = 10.508 mH \tag{10.7}$$

The integer is 51 turns. The A_L value is 4040 using F material as before. This is 2.5 mH more than our goal of 8 mH. Solve for the turns needed with the normal A_L formula.

$$N_t = 1000\sqrt{\frac{8}{4040}} = 44.499 \tag{10.8}$$

Use the upper integer, or 45 turns, for each half of the windings. Keep the end gaps between the two halves as far apart as possible. This creates a visible winding gap on the core and makes this gap as large as can be. The difference between 51.88 and 45 gives the gap spacing between both ends of the two windings; 6.88 times the wire diameter of 0.0429 in. gives approximately this spacing. This is on an arc, and the whole turn will not touch, thereby eating some of this circumference; so the value will be less than the 0.295 in. calculated. Again, either tape or fill these spaces so that both gaps between the two windings are secure. If this is not enough, redo with a next-best core. Now the next question: Will it fit the required box? Add 2.2 times the wire diameter to the O_D to get the outside diameter, which may not fit the box. Try two smaller cores stacked together; the A_L value doubles.

Now that the method of obtaining the proper core size is known, how was the value of 8 mH determined for the Zorro? Two things must be resolved. The first is how to convert from the common mode to the differential mode—really, from balanced (as the common mode is) to unbalanced, often called normal mode. If several balanced normal-mode networks follow the common-mode inductor, this must be converted back to unbalanced differential mode. All this is done to ease the calculations.

In Figure 10.2, the *Z* is first followed by two L filters in turn, and then followed by feed-through capacitors. In the initial differential calculations, a double L filter was required, and two differential-mode inductors were calculated to provide the value of L1 for each inductor. To switch to balanced mode, L1 was divided between the two legs shown in Figure 10.2 as 0.5 L1. In common mode, these two 0.5 L1 inductors are in parallel and equal 0.25 L1 in common mode. The C1 capacitors are out of the circuit. These four inductors equate to 0.5 L1 common mode and add to the 8 mH. Usually, the 0.5 L1 value is so small compared to the 8-mH Zorro that they can be neglected. This converts to a single inductor of 8-mH value, and the two feed-through 0.02-μF capacitors are doubled in value. The reason is that the total value of these two coils may be around 400 μH, which is an order of magnitude lower than for the Zorro inductor. The two feed-through capacitors may be limited in size by the specification, such as MIL-STD-461, where the maximum for 400 Hz is 0.02 μF but now totals 0.04 μF. This is now in a form where the actual value of the common-mode inductor can be solved using techniques discussed in chapters 16 and 17. The next point to discuss is the test setup used to test the common-mode inductor, which is shown in Figure 10.2. The two input leads are shorted together. The input is fed from a tracking generator with 50 ohms output impedance. The two filter output leads are also shorted together and feed the load, which is the analyzer and is also 50 ohms. This was another reason to convert from the balanced to the unbalanced filter as before, because the common-mode inductor can be calculated via the equations in chapters 16 and 17.

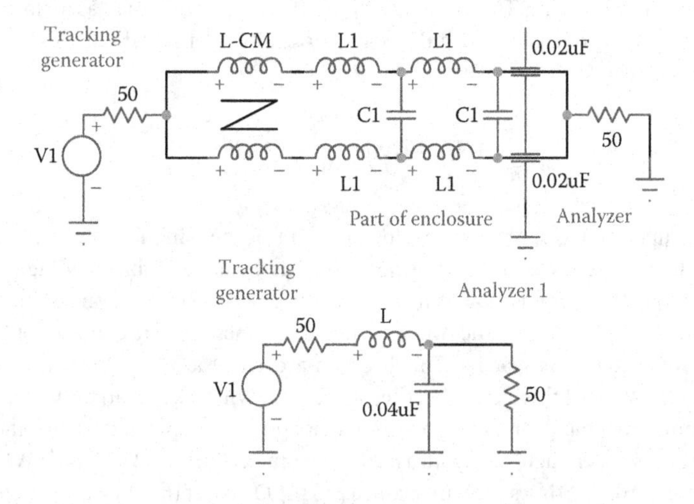

FIGURE 10.2 Testing a balanced common-mode filter and equivalent circuit.

10.5 Common-Mode Calculation

The method used to arrive at Figure 10.2 for the equivalent circuit is as follows. C is equal to two times C2, L is the common-mode inductor needed for the proper insertion loss, R_S is the source impedance, and R_L is the load impedance. In Figure 10.2 both source and load are the same value at 50 ohms. Because the equation includes R, L, and C, equations based on *charge* ($Q = CV$) rather than I (current) of the two networks are the easiest to work with, and this generates the matrix in equation (10.11). This is from impedance matrix equations.

$$\begin{bmatrix} V_i \\ 0 \end{bmatrix} = \begin{bmatrix} Ls^2 + Rs + \frac{1}{C} & -\frac{1}{C} \\ -\frac{1}{C} & Rs + \frac{1}{C} \end{bmatrix} \begin{bmatrix} Q_1 \\ Q_2 \end{bmatrix} \tag{10.9}$$

The delta determinant of the matrix is as follows

$$\Delta = LCRs\left[s^2 + \frac{(R^2C + L)s}{LCR} + \frac{2}{LC}\right] \tag{10.10}$$

Substitute in the initial requirements of V_i and 0 and solve for Q_2

$$\begin{bmatrix} Ls^2 + Rs + \frac{1}{C} & V_i \\ -\frac{1}{C} & 0 \end{bmatrix} \tag{10.11}$$

Solve for Q_2

$$Q_2 = \frac{V_i}{C}\left[\frac{1}{LCRs\left[s^2 + \frac{(R^2C + L)}{LCR}s + \frac{2}{LC}\right]}\right] \tag{10.12}$$

Note that Q_2 is not the goal; V_0 is required. In reality, the goal is the ratio between output and input voltage.

$$Q_2(s) = I_2 \qquad I_2R = V_0$$

$$\frac{V_0}{V_i} = \frac{1}{LC^2\left[s^2 + \frac{(R^2C + L)}{LCR}s + \frac{2}{LC}\right]} \tag{10.13}$$

This equation has been published in many articles, but most often they do not include R_S, the source impedance. In many cases, they do not use the two feed-through capacitors in parallel that are now doubled in value, but instead solve by completing the square. This means, in the case of Figure 10.2 and Equation 10.15,

$$a^2 = \left[\frac{R^2C + L}{2LCR}\right]^2 \tag{10.14}$$

that a^2 must be added to complete the square of the s terms in the main denominator and subtracted from the last term, being $2/LC$. The a^2 term is always much greater in the common-mode application than the last term in the main denominator, making the new last term, $-\omega^2$. This makes the solution a hyperbolic function and very lossy, as suggested by the test setup. Typically, when performing this analysis, R_S is left out of the equation, making the value of a reduce to $1/2RC$, the damping factor. This also reduces the last term of the denominator to $[1/(LC) - a^2]$.

$$\alpha = \left|\frac{(R_s R_L C + L)}{2LCR_1}\right|$$

$$\omega = -\sqrt{\left|\frac{(R_s R_L C + L)}{2LCR_1}\right|^2 - \frac{(R_s + R_1)}{LCR_1}} \tag{10.15}$$

This defines a, and is the damping factor.

$$a = \frac{1}{LC}\left|e^{-at}\sinh(\omega t)\right| \tag{10.16}$$

This may appear to be simplistic. However, there is another way by knowing that the common mode will always be lossy. Forget the sin or sinh solution and define the main denominator of equation (10.13) to be a quadratic form

$$D(s) = (s + a)(s + b) = s^2 + (a + b)s + ab \triangleq s^2 + 2\zeta\omega s + \omega^2 \tag{10.17}$$

so that $a + b = 2\zeta\omega s$ and $ab = \omega^2$ of the main denominator of equation (10.13) is repeated here.

$$\frac{V_0}{V_I} = \frac{1}{LC^2\left[s^2 + \frac{(R^2C + L)s}{LCR} + \frac{2}{LC}\right]} \tag{10.18}$$

This is a simple solution, and both a and b are included within the same quadratic. Either a or b can be assigned the positive square root, but the solution is better with b being the more positive

$$a = \frac{(R^2C+L)-\sqrt{(R^2C+L)^2-8LCR^2}}{2LCR}$$
$$b = \frac{(R^2C+L)+\sqrt{(R^2C+L)^2-8LCR^2}}{2LCR} \tag{10.19}$$

and takes the form

$$\frac{e^{-at}-e^{-bt}}{(b-a)} \tag{10.20}$$

The term $b-a$ reduces to

$$b-a = \frac{\sqrt{(R^2C+L)^2-8LCR^2}}{LCR} \tag{10.21}$$

This cancels LC, so the final answer to V_o/V_i is given in equation (10.22)

$$\left|\frac{V_o}{V_i}\right| = \frac{R\left|e^{-at}-e^{-bt}\right|}{\sqrt{(R^2C+L)^2-8LCR^2}} \rightarrow 20\log_{10}\left|\frac{V_o}{V_i}\right| = dB_{\text{LOSS}} \tag{10.22}$$

See chapters 16 and 17 for a discussion of other methods. In the normal test arrangement, R is 50 ohms and C, because of leakage current specification, is whatever the specification requires. Remember, the value of the capacitor is doubled here because the two feed-through capacitors are in parallel and, therefore, they add. Now that the common mode is reduced to a simple single L filter, the required value of the common-mode inductor can be easily solved, as seen in chapters 16 and 17.

10.6 Differential Inductance from a Common-Mode Inductor

Some filter manufacturers have designed common-mode inductors that also function partially as differential-mode inductors. As before, this is done with very wide winding spacings that generate the leakage inductance needed to provide this differential inductance. Some flux of one coil fails to cut some of the other windings, creating this leakage inductance. Another way this is accomplished is by using pot cores. These use a split bobbin so that some of the flux in one half of the bobbin fails to cut the other half. This has been expanded toward using two separate bobbins that fit in the core with additional room to place a washer between the two bobbins. This washer is cut, or split, if it is a conductor to avoid the washer acting as a shorted turn. The material of the washer has little to do with the differential inductance created. It is the separation of the two windings that causes the leakage inductance. If the spacer is Mylar, the washer does not have to be cut. The leakage inductance is easy to measure with an inductance bridge. Shunt one winding of the common-mode inductor, and read the inductance of the other winding. If all the flux lines cross or cut the other turns, the reading is zero. This is truly impossible to accomplish because the turns cannot all be so tightly coupled. The difference is the leakage inductance measured by the inductance bridge. Another way is to measure both legs together, opposing, and the inductance bridge will read the leakage inductance. Some people suggest that the leakage inductance is due to flux leakage in air and not the core, so it cannot saturate. This is true to some degree, but it saturates a minor amount because not all of the flux is in the air.

10.7 Common-Mode Currents—Do They All Balance?

In some cases, engineers think that common-mode inductors require a balanced sine wave for the common-mode source to work properly. In the single-phase common-mode inductor, the sum of the magnetic fields caused by the currents will still cancel no matter what the wave shape is. Because the current is equal and opposite, these two flux fields still cancel. If the currents are not equal—for example, a current difference created by the capacitors to ground—the two fields almost cancel and give some differential-mode inductance. This is the problem in speaking to purists, most of whom claim that the feed-through capacitance unbalances the common mode because they are not matched. All of this is somewhat true, but to what degree? This is also true of any arresters from line to ground because they have different values of capacitance across them and leakage currents through them.

This statement is very true in three-phase circuitry where the phases are exactly 120 degrees apart. Assuming this is not happening, the wave shape still need not be sinusoidal. Any current of any shape gives a magnetic field that cancels in the single-phase, three-phase, or DC system. If the reverse was true, common mode would not work for three phase. Because of the imbalance of the phases and the harmonic content, the voltage is not very sinusoidal; thus, the common mode will not work. The three-phase Zorro requires three windings for the delta and four wires for the added neutral in the wye. If the three legs are not balanced, the neutral carries the difference current and magnetic flux will still cancel because of the fourth neutral winding.

11 Transformer's Addition to the EMI Filter

The transformer is overlooked as an EMI filter element. This is because of the obvious disadvantages of weight, size, and cost. Using off-line regulators and switchers can often eliminate transformers. On the other hand, switchers often complicate EMI issues because of the high-frequency switching noise they generate. This is worse as the switcher frequency increases. Also, switchers, because of their present-day glut of components, may have a tendency to impact MTBF (mean time before failure) requirements. A question for the system designer is perhaps, "Do the disadvantages of using a transformer overshadow the advantages?"

11.1 Transformer Advantages

The main advantages of the transformer, without considering their EMI advantages, are isolation, voltage translation, common-mode rejection, and the potential of low-leakage current. Another advantage that is overlooked is the ruggedness of this device. Transformers can handle voltage spikes without difficulty. If the transformer is an autotransformer, isolation, common-mode rejection, and leakage-current advantages are done away with because of electrical cross-coupling of the secondary to the primary. This chapter assumes that a standard isolation transformer is used.

11.2 Isolation

Isolation is accomplished because the primary and secondary are coupled magnetically rather than physically or tied together electrically. Therefore, whatever the wiring arrangement of the secondary is, such as one output leg tied to chassis ground, it is not coupled electrically to the primary. The green safety chassis ground lead does not carry any current back to the service ground from the secondary, even though the secondary is tied to this chassis ground.

11.3 Leakage Current

There are various specifications regarding leakage current. In commercial applications, the leakage current specification is typically 5 mA for the system. The easiest medical requirement is 300 μA (0.3 mA) for the system, and requires even tighter specifications if the unit is attached to the patient. The tighter medical specification for the transformer is usually specified as 100 μA. Leakage current through the transformer is, in theory, zero, but in reality it is a function of spacing and shielding between the primary and secondary. The leakage-current reduction is also a function of a Faraday screen, if used. For transformers with bifilar winding, where the primary and secondary are wound together at the same time, the leakage current is drastically higher. The reason for all this is that the leakage current is a direct function of the primary-to-secondary capacitance. If the windings are bifilar, there is little spacing other than the wire insulation. Obviously, no Faraday screen or shield can be used, so the leakage current is due to the capacitance from wire to wire and may very well be higher than anticipated or desired.

11.4 Common Mode

Common mode tracks directly with the leakage current of the transformer. Common mode—equal voltage on both lines of either the primary or the secondary—does not create a magnetic field across the primary or secondary of the transformer. The only coupling is through the capacitance from primary to secondary. Any reduction in the leakage current by reducing the capacitance also reduces the common mode. One cure for this is the Faraday screen that is located between the primary and secondary, but research years ago showed that this was a function of frequency: The higher frequencies were coupled across from primary to secondary. The other way was to provide spacing between the primary and secondary. This way, the leakage could be reduced below 50 μA and also reduce the common mode.

11.5 Voltage Translation—Step Up or Down

Step-up and step-down requirements within transformers were the main reason for their development years ago and, therefore, are not discussed here.

11.6 Transformer as a Key Component of the EMI Package

The problem with EMI filter design is opposing the stated requirements. One of these is a heavy common-mode requirement specified along with an impossible leakage-current requirement. These two items conflict because, without a transformer, the filter requires sizable capacitors to ground on both the hot and the return wires and a common-mode inductor to remove the common-mode noise. The leakage current requires little or no capacity to ground. A quality transformer can often solve all these technical issues. The transformer provides the filter designer with both common-mode and differential-mode loss. If the EMI filter is first in the power stream, care must still be taken in the capacitors-to-ground values for leakage requirements. However, the transformer should have removed most of the common-mode noise anyway, easing the filter needs for the common-mode noise requirement. So, small capacitors and common-mode inductors should be sufficient to do the job. If the EMI filter follows the transformer, any reasonable capacitor values could be used because the transformer will eliminate this ground current from the primary side, as this current is common mode.

We ask ourselves, "Is this all the transformer would do for the filter designer?" Certainly not. Power transformers using laminations, C cores, and tape-wound toroids with steel thicknesses of 12, 11, 7, and 4 mils exhibit very high wattage losses in watts per pound at higher frequencies. This enhances the differential-mode section of the transformer. These graphs are hard to read, but MK Magnetics, located in Adelanto, California, provided the Arnold 12 Mil Selectron 'C' Core Watt per Pound graph (Figure 11.1). With the graph, the loss per octave and decade can be closely approximated for 900 gauss, as shown in Table 11.1.

$$\begin{aligned}
(a) &= 10\log\frac{0.27}{1.1} = 6.1 \\
(b) &= 10\log\frac{24}{93} = 5.8 \\
(c) &= 10\log\frac{0.27}{24} = 19.5 \\
(d) &= 10\log\frac{1.1}{92} = 19.22
\end{aligned} \tag{11.1}$$

In these equations, the weight in pounds cancels, and the flux density in gauss was a constant, leaving watts divided by watts. In equation (11.1) a and b are for the octave (frequency doubling) loss using the data from table 11.1. Both are near 6 dB per octave.

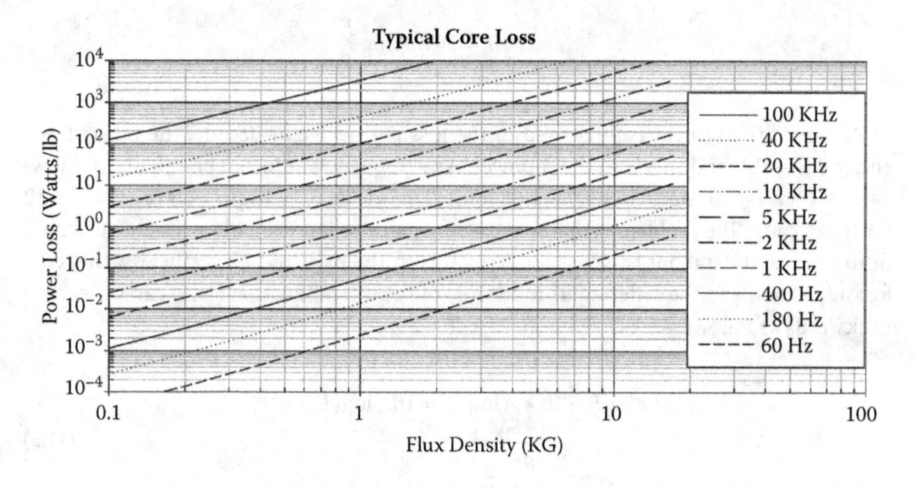

FIGURE 11.1 Watts per pound chart at 900 gauss.

TABLE 11.1 dB Calculations of Loss per Octave and Decade for 900 gauss

Hertz	Watts per pound
1,000	0.27
2,000	1.1
10,000	24
20,000	92

Note: This table was derived from the graph in Figure 11.1.

The next two (c and d) are for the decade (10 times the frequency) loss, which gives 20 dB per decade. This loss, or cutoff frequency, starts near the fifth harmonic for the steel type. For steel, 12 mil is proper for 60 Hz, and this octave, or decade, loss should start by 300 Hz. This means that the tester should expect to see something close to 6 dB by 360 Hz and close to 20 dB by 1,800 Hz. However, the cutoff frequency would vary from transformer to transformer. This differential-mode loss is dissipated, not attenuated. Another way to evaluate this is through the core manufacture estimated-loss equations, as seen in equation (11.2). These have the form Watt/lb = CF^AB^E, where C is a constant (possibly to a power), F is the frequency at some power A, and B is the flux density at some power E. Because the weight (lb), flux density (B), and constant C of the core remain the same, they cancel. Therefore the standard dB equation can be used.

$$\frac{\text{Watt}(1)/\text{lb}}{\text{Watt}(2)/\text{lb}} = \frac{C\ F_1^A\ B^E}{C\ F_2^A\ B^E}$$

$$\frac{\text{Watt}(1)}{\text{Watt}(2)} = \frac{F_1^A}{F_2^A} = \left(\frac{F_1}{F_2}\right)^A \tag{11.2}$$

$$dB = 10 \times A \log \frac{F_1}{F_2}$$

Because the engineer is interested in the loss per octave or decade, the ratio of F is either 0.5 or 0.1, and this is to the A power. Any frequency ratio can be calculated; however, Armco 14-mil steel is listed as 1.68 for A. This yields 5.05 dB per octave and 16.8 dB for the decade. The problem is that with these equations it is difficult at best to get a near fit to the listed data, but this loss for Armco is in the ballpark. Nevertheless, this core loss adds greatly to the differential-mode loss of the filter. Equation (11.2) can be worked backwards as follows:

$$dB = 10 \times A \log \frac{F_1}{F_2} = 10A \log(F_R)$$

$$A = \frac{dB}{10 \log(F_R)} \tag{11.3}$$

where F_R is the frequency ratio, which is 0.5 for 6 dB. Back-solving for A, with a dB of 6 and F_R equal to 0.5, gives 1.993 for A. As an example, say a transformer has 6 dB of loss at 600 Hz. The designer needs 60 dB at 20 kHz. The frequency ratio 20,000/600 equals 33.33. The logarithm of 33.33 to the 1.993 power is 3.035, and this times 10 is 30.35 dB. Adding back the 6 equates to 36 dB. The designer needs 60 dB at 20 kHz, so 24 dB more loss is needed. A double-L filter added to the system would do nicely. Four elements at 6 dB each give 24 dB. The cutoff frequency for the double L would, in theory, be 10 kHz. These would prove to be reasonably small components giving very high self-resonant frequencies (SRFs). Is there any other action within the transformer to aid in the loss?

11.7 Skin Effect

The higher frequencies of either common or differential mode are also dissipated within the high-frequency resistance of the wires. However, this does not come into play until approximately 30 kHz and above. The radius of conduction, in centimeters, is

$$R_C = \frac{6.62}{\sqrt{F}} \tag{11.4}$$

11.8 Review

In other words, the isolation transformer adds the same as any other EMI filter element—6 dB per octave or 20 dB per decade. A further advantage is that the frequency cutoff point is so much lower than for the normal EMI filter components. Heavy common- and differential-mode loss is realized by using the isolation transformer. Again, disadvantages are the added weight, size, and possibly cost. If the transformer eliminates one or more filter sections, the increase in cost may be compensated by eliminating the costs of these components. For example, if a 12-mil (for 60 Hz) steel core gives only 6 dB at 600 Hz, an additional 20 dB by 6,000 Hz, and another 6 dB by 12,000 Hz, the total loss is 32 dB at 12,000 Hz, and we can add a few for 14,000 Hz. This may be more than enough loss so that no other filter elements may be needed.

On the other hand, the transformer's effectiveness diminishes as the primary-to-secondary equivalent capacitance comes into play. The approximate primary-to-secondary equivalent capacitance is

$$C = \frac{I}{2E\pi F} = \frac{100 \times 10^{-6}}{2 \times 120 \times \pi \times 60} = 2210\text{pF} \tag{11.5}$$

For the purposes of example, E is the input volts at 120, F is 60 Hz, and I is 100 μA. The maximum value would be 2,210 pF, and 50 μA would be half that value. In a 220-A test setup, the transformer would be effective to approximately 700 kHz for well-designed and manufactured transformers. This is where the capacitor would be equal to the source plus load impedance, in this case, 100 ohms. However, the transformer effectiveness would be much lower than this frequency because of the SRF caused by leakage inductance. But then again, all filter components suffer from this dilemma, not just the transformer.

12

Electromagnetic Pulse and Voltage Transients

This chapter develops an impedance ratio according to the following ideas. The circuit discussed here consists of a battery, a transmission line, and a switch. The far end of the transmission line is either a short or an open. The battery has an output impedance, Z, that equals the impedance of the transmission line. When the switch is closed, a step function travels down the line at the velocity of propagation. The voltage divides between the characteristic impedance of the line and the source impedance. The main interest to us is related to how the far end is terminated, or not, as the case may be. If the line is open at the far end, the voltage doubles. The pulse energy travels back at the same velocity and elevates each segment of the transmission line to full voltage. At the time the wave reaches the source, or battery end, the current drops to zero, and the full battery voltage is impressed across the transmission line (Figure 12.1).

If a short is at the far end instead of an open, the current doubles when the wave reaches the shorted end. One half of this current depletes the initial stored voltage, segment by segment, while the other half continues to flow through the line from the battery. This discharges each segment. When the pulse reaches the battery end, the line is fully discharged, and double current flows from the battery, or source. The full voltage is dropped across the internal source impedance, Z. If the line and battery source impedance are not equal, the pulses iterate back and forth until equality is reached (reflection coefficient). The main point here is to note the difference between the two far-end conditions: one near open and one near short. In the link, or short line, the source impedance is thought to be very low and, again, the question is: What is at the filter end of the line? Is it very low impedance or a short circuit condition; or is it a high impedance, or nearly open, condition? If the initial condition is high impedance for the load, the full-strike voltage is impressed across the suppressor, aiding turn-on or firing. The device then drops to low impedance, and a high current is carried through the suppressor. The suppressor and the line impedance will dissipate the energy—we hope. If the energy is too great for the suppressor, the suppressor normally first shorts and then blows open.

The power in the pulse is dissipated in the TVS, line, and source impedance. If the condition is near short, the voltage divides, according to the impedance ratio, and delays the firing of the protector. This is similarly displayed in Figure 12.2. Some engineers

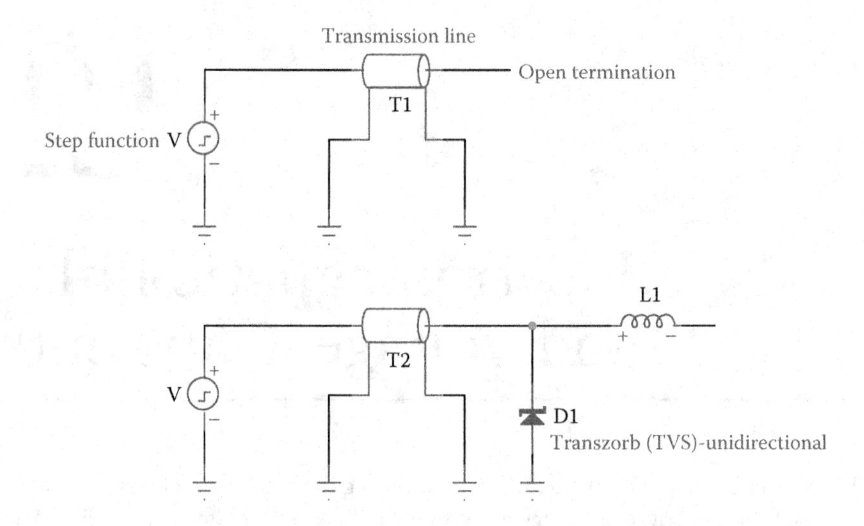

FIGURE 12.1 Open-ended transmission line applied to a Transzorb (TVS).

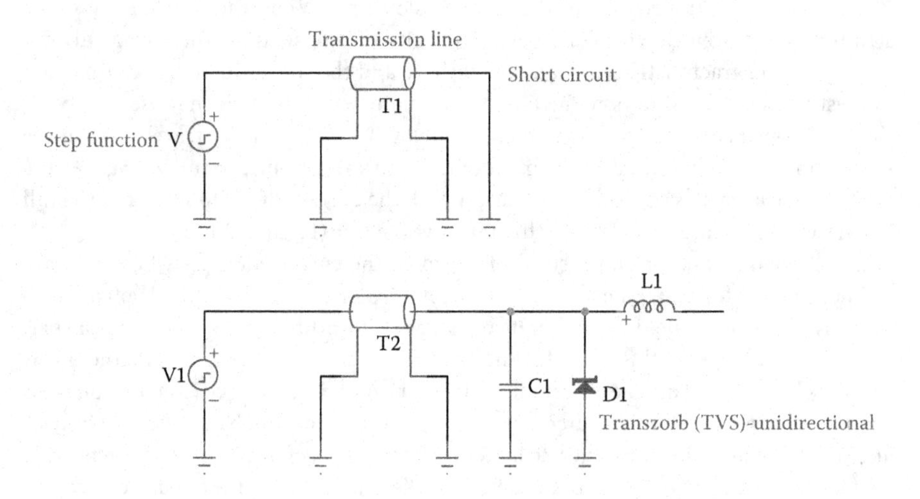

FIGURE 12.2 Shorted transmission line applied to Transzorb.

place capacitors across the TVS, making the condition similar to the preceding shorted condition. The turn-on time for TVS devices and MOVs are much faster these days, but the capacitor must charge to well past the turn-on voltage before the arrester can act. Another point is that the pulse current through the capacitor is on the order of twice the initial line current or several times 100 A. This action delays the turn-on of the TVS; therefore, the transient stress seen by the components downstream of the TVS will be much higher.

This event may also lead to capacitor failure if it was not rated for this amount of pulse current regardless of its construction, and again if the capacitor voltage rating was exceeded. This is especially true if the capacitor is the metallized film type. It is argued that placing a capacitor in parallel with the TVS will aid in protection; however, when a high-energy pulse occurs, both the capacitor and TVS can fail. If the far end (filter input end) is open—TVS directly across the line followed by an input series inductor—the voltage will rise quickly because the inductor acts as an open (high impedance). The voltage will not necessarily double; the impedance ratio is not known. The quickly rising voltage will help the arrester to fire, speeding up turn-on. The Transzorb may blow, but the filter and the following equipment will be protected. Most of these arresters conduct in less than a microsecond under these conditions, so the peak current through the inductor and the voltage stored across the following filter capacitor are reduced. To summarize, it is recommended that the arrester (TVS or MOV) be placed at the input to the filter inductor and without a parallel capacitor, so that the impedance to a high-energy pulse is high. This will result in a very fast rise in voltage, and the arrester will turn on fast enough, thereby protecting the rest of the circuit.

12.1 Unidirectional versus Bidirectional

The selection criteria for the type of arrester, e.g., Transzorbs or TVS diode and MOVs, are somewhat driven by the circuit application and will need to be carefully selected based upon the input power architecture, including any lightning or transient voltage requirements.

In the case of a Transzorb (TVS), the application may require either a unidirectional or bidirectional device of a particular breakdown voltage (V_{BR}), clamping voltage (V_{CL}), and power handling capability. For DC applications, a unidirectional device may be used, as this will clamp positive pulses, or spikes, in avalanche mode, while negative spikes are clamped by the diodes' forward conduction. Bidirectional arresting devices may be used in both floating, or differential, DC power applications and in AC applications.

Transzorbs are typically selected by reviewing the input operating voltage for the equipment, the transient energy that needs to be dissipated, and the circuit's absolute maximum voltage rating requirements. The breakdown voltage, V_{BR}, must be greater than the supply or signal voltage, including any tolerance variation. The TVS wattage rating must be capable of withstanding the transient energy. The clamping voltage, V_{CL}, must be lower than the absolute maximum voltage of the circuit and its components to provide protection against overvoltages. To improve signal attenuation or insertion losses, a low-capacitance device should be selected. This should also reinforce the concept that placing capacitors in parallel with TVS devices will slow the response time of the TVS itself, thereby creating a higher potential for stress on downstream components as well as excessive losses in the circuit. The goal is to ensure that the arrester is forced into conduction as fast as possible.

12.2 Three Theories

The first theory comes from those who say that the arrester is not necessary and that it is the job of the filter to handle these high pulses of energy. This might be true if the components were designed to handle them. In this case, the inductor must be designed to handle the full pulse voltage and pulse duration without arcing. The following capacitor must withstand twice the pulse voltage without failing. Most wire insulation withstands 500 V and should not require special insulation. If the initial surge voltage is over 2,500 V, the turns of this special inductor should not touch each other. The insulation withstand voltage is easy to measure. Take the length required for the inductor and strip one end. Bury this length of wire with both ends exposed in an aluminum basket of shot pellets, or any other small conductive container. Apply the test potential to the stripped wire end and the other high-voltage lead to the shot container. To try to comply with the first theory, the initial inductor would have to sustain the initial voltage. The problem is the inductance value. A spring-type coil still needs a reasonable value of inductance. In equation (12.1), T is the length and R is the radius in inches; L is in μH, where N is the number of turns. Note that as the coil is stretched out, L drops; however, as the radius increases, the value of L increases. It is necessary to spread the turns out for the high-voltage pulse and push them together for a higher L value.

$$L = \frac{0.8\mu_{eff}(RN)^2}{(9R+10T)} \tag{12.1}$$

12.3 Initial High-Voltage Inductor

In equation (12.1), μ_{eff} is the permeability of air, which is 1. Increase the radius, if possible. However, the wire gets longer and harder to keep the shape without an insulated tubelike glass to wind it on. As in Figure 12.3, the turns spacing can drop as the turns are wound. The inductor acts as a transmission line, and the full spike voltage is impressed across the first turn, one at a time. This charges the capacity to ground for each turn and then moves on to the next turn. Many inductors impressed with high voltage seem to burn out on the first few turns, and this dictates the winding spacing. Figure 12.3 is the small inductor in front of the arrester followed by another inductor, which is the first filter element. This first inductor widens the pulse width and drops the peak current—the same energy spread over a longer time and a lower current peak.

The second theory considers protecting the equipment and not the filter, so the arresters are placed at the equipment end of the filter. It is also argued that, perhaps, we may not want to exercise the arresters on every pulse that is seen at the input. This is a valid point because these units will only take so many hits before they are likely to fail. The questions

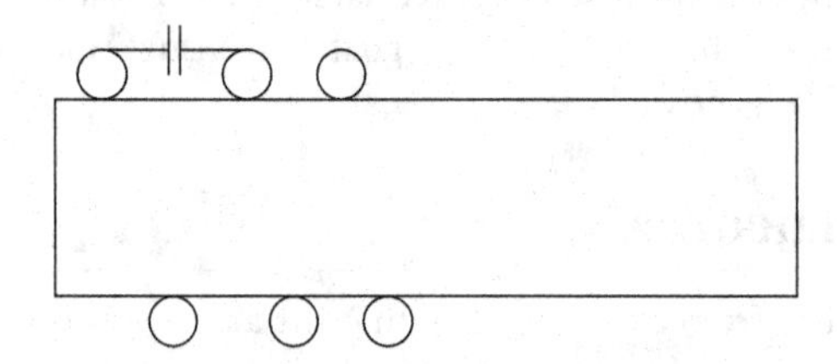

FIGURE 12.3 Small inductor in front of the arrester.

asked are: "What if the pulse destroys the filter?" and "Can the equipment still operate?" These valid questions are indeed answered with a question. If the arrester fails, can the equipment still operate? The answer to the question could be yes for two reasons.

1. Through the action of the arrester closing and providing a low-impedance path for the fault current, the arresters often fail short circuit, then with the high current through them, they eventually fail open. If the arrester is then open, the equipment can still function.
2. The best place for the arrester is at the input to the filter, positioned between the line input and the filter inductor (could be first of several) so that the pulse will see high impedance. With proper access to the filter, a failed arrester, if practicable, may be removed, and the equipment can still functionally operate until new arresters can be installed.

The third theory considers locating the arresters at the front end of the filter to protect the entire equipment. The diatribe of the opening comments also pertains to this solution. The arrester (MOV or Transzorb/TVS) is located at the front end with an inductive input to follow as part of the filter such as shown in Figure 12.1. When the pulse reaches the arrester and inductor, the pulse sees high impedance from the inductor; the voltage rises rapidly, quickly firing the arrester. The initial line current continues through the arrester, and the pulse is dissipated in the arrester, the line impedance, and the skin effect in the wiring. The voltage that the filter sees is the arrester clamping voltage. In addition to this, any capacitors that follow the inductor should be capable of withstanding at least twice the MOV or Transzorb clamping voltage. There is a fourth rationale that is a combination of theories one and three: The input inductor is split, with the arrester tied to the junction of the inductors. The first half limits the arrester current but must be able to withstand the pulse. This first inductor is often wound on a bobbin without a core to eliminate the potential for any arcing to the core as shown in Figure 12.3. This inductor has the effect of both slowing and widening the pulse, thereby reducing the peak voltage.

Finally, another solution might use two arresters, one on each end. This is for very high pulses on the order of 100 kV. Often the arrester is sandwiched between two input inductors, and the last is across the output capacitor. The filter will take the remaining energy not handled by the initial arrester and spread it over time, lengthening the pulse width but reducing the peak energy. The last arrester will handle the remaining energy.

12.4 Arrester Location

From the preceding section, it should be clear that the preferred method is to place the arrester at the input to the filter, with the filter having a series inductor at the input. For EMI filters that are manufactured in closed boxes, the preferred method is to mount them outside the filter, where there should be access to replace them if the need ever arises. For a PCB design, placing them at the input to the filter ensures that the copper

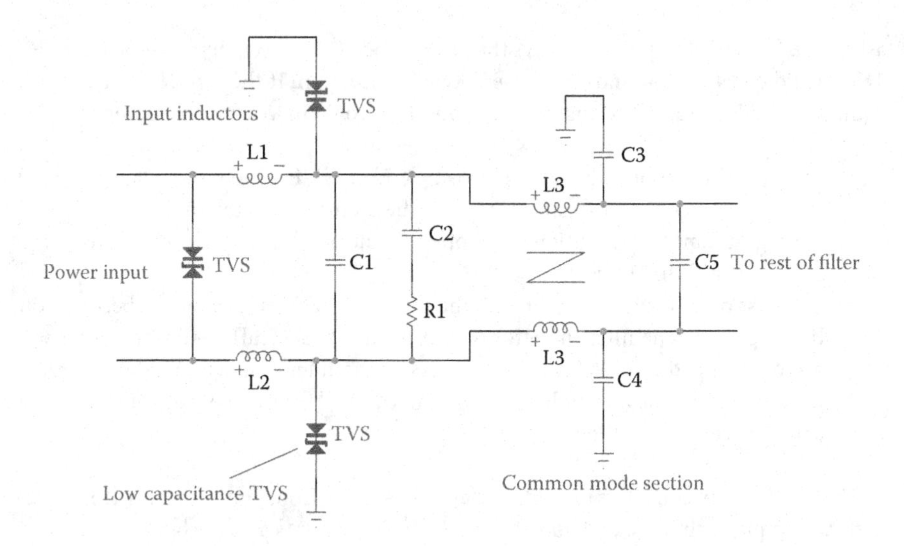

FIGURE 12.4 Common-mode and differential-mode arresters at the input.

traces, both to and from the arresters, are large enough (cross-sectional area) to withstand the pulse current without damaging the copper. Typically, if the filter is designed to carry high current, especially for low-voltage DC systems, then the copper traces should already be sized to handle the arrester short-circuit current.

In the single-phase balanced circuit, three TVS devices are required: one from hot to ground, one from neutral to ground, and one from line to line. The common-mode TVS devices protect the equipment from common-mode pulses—pulses from both lines to ground. In Figure 12.4, they are shown on the common-mode side of the input inductors. This is done to reduce the size of the TVS, as the inductor will help to limit the initial current pulse. The TVS connected line to line provides differential-mode protection at the input of the filter. If the common-mode TVS devices are at the input to the filter, they will need to be sized to handle the peak pulse power and current.

In the three-phase filter, where all three lines are treated within the one filter enclosure, six TVS devices are required. Three are wired from the three lines to ground, and three are wired from line to line. If the units have internal arresters, the entire filter must be replaced after an event such as EMP or other pulse that destroys the TVS. This can create an inventory problem requiring expensive replacement of filters or PCB assemblies. Where possible, the arresters should be located where they can easily be replaced.

12.5 How to Calculate the Arrester

The specification of the shape and size of the strike is usually specified within the system or design requirements. Thereafter, it is possible to calculate the joules required for dissipation, the type of TVS device, and the clamp voltage required for the application to ensure adequate protection. Knowing the waveform and peak voltage, current, etc., the joule rating of the TVS may be calculated.

12.5.1 Dynamic Resistance

The primary goal for using an arrester or TVS is to provide the lowest resistance shunt path to ground for the pulse current. It is important to recognize that a TVS is primarily intended to serve as a shunt-voltage clamp across the input of the filter to protect sensitive components in the circuit from high-voltage transients. Until these transients occur, the TVS will be idling at very low standby current levels and appear "transparent" to the circuit. When a high-voltage transient does occur, the device clamps the voltage by avalanche breakdown. So, the TVS is really a variable resistor and will be high impedance during normal circuit operation and low impedance during an electrical overstress event. The dynamic resistance may be calculated by dividing the clamping voltage by the peak impulse current.

$$R_{dynamic} = \frac{V_{CL}}{I_{PP}} \Omega \tag{12.2}$$

A method of determining the selection of a TVS follows. We assume that a bidirectional TVS is connected across the input of a DC power input with a maximum input voltage of 100 V. This voltage will determine the rated standoff voltage (V_{WM}) selection of the TVS device where, under normal voltage conditions, the impedance is high with low standby current. We now select the next voltage level, which is the breakdown voltage (V_{BR}) at $V_{WM}(1.2) = 120$ V. A tolerance factor of 20% is used as a safe margin and is also in place to allow for the device temperature coefficient, which can be in the range of ±0.1%/°C. The breakdown voltage is where device operation transitions from a high-impedance standby condition to avalanche, and is where the impedance starts to rapidly fall. Finally, the maximum clamp voltage (V_{CL}) is selected, and this is typically 30% to 40% higher than the breakdown voltage (V_{BR}). If we have a breakdown voltage of 120 V and the maximum clamping voltage is 40% higher, we select $V_{CL} = 120(1.4) = 168$ V. For the application, we select a TVS with a breakdown voltage of 126.5 V, which has a maximum clamping voltage of 160 V at a peak impulse current (I_{PP}) of 100A. As we have previously noted, the components in the filter, namely the capacitors, must be rated at least 20% higher in voltage than the breakdown voltage to ensure component protection during an overvoltage event. In this case, we would use 200-V capacitors.

All TVS devices are rated in various peak pulse power dissipation (P_{PP}) levels so that a variety of surge conditions can be managed safely. To select a TVS in P_{PP} by calculated methods, it is necessary to define the transient conditions in both peak impulse current (I_{PP}), pulse width, and waveform. Most often, I_{PP} is specified with a 10/1000 μs waveform similar to that of Figure 12.10. The maximum peak pulse current rating (IPP), defines the maximum current handling capability for a given pulse duration. The current handling capability is referenced to a specific wave shape (i.e., 8/20μs, 10/1000μs) and is not constant over time. The P_{PP} is the product of the maximum clamping voltage multiplied by the peak impulse current for a given waveform and pulse width, or

$$P_{PP} = V_{CL} I_{PP} \tag{12.3}$$

We now consider the shape of the waveform where each has a unique K factor. This shape factor, K, is needed to define the joule rating requirement. See Figures 12.5–12.7.

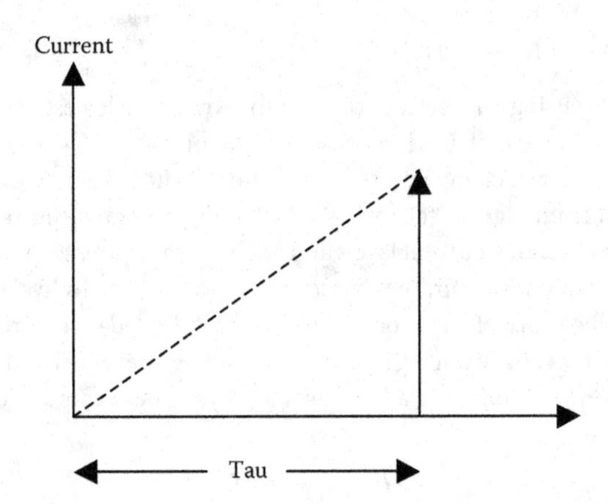

FIGURE 12.5 The ramp, $K = 0.5$.

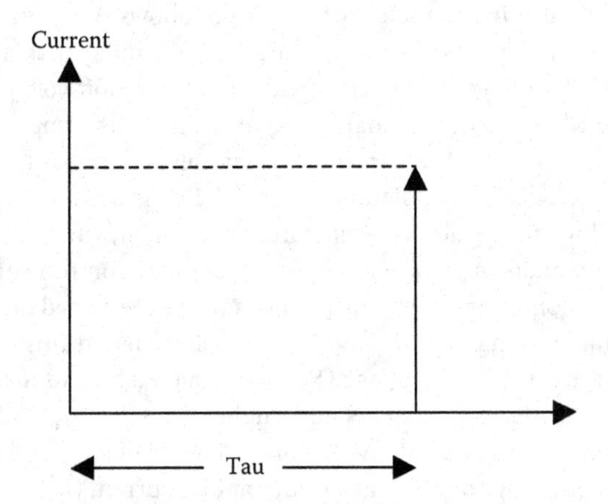

FIGURE 12.6 Constant height, $K = 1.00$.

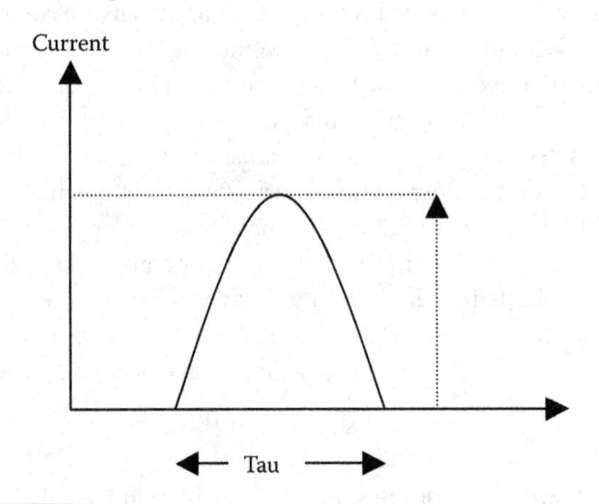

FIGURE 12.7 Sign pulse, $K = 0.637$.

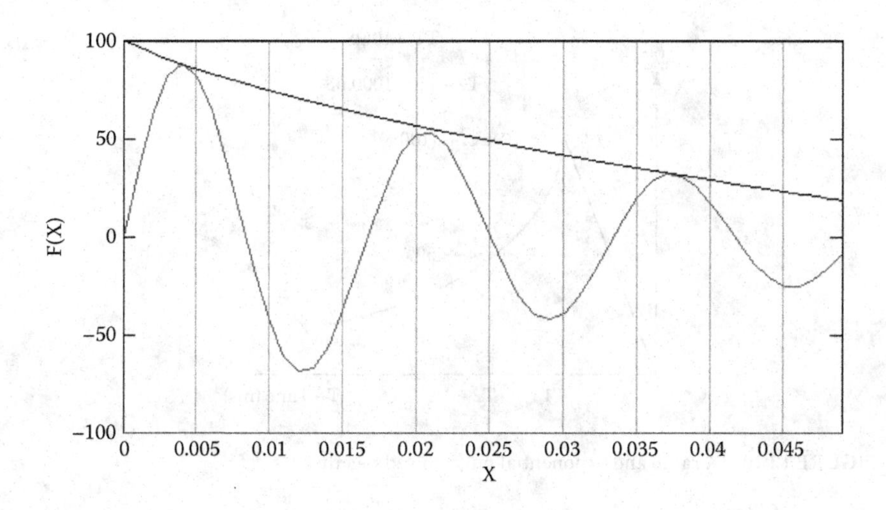

FIGURE 12.8 The dampened sine, tau based on 50% current, $K = 0.86$.

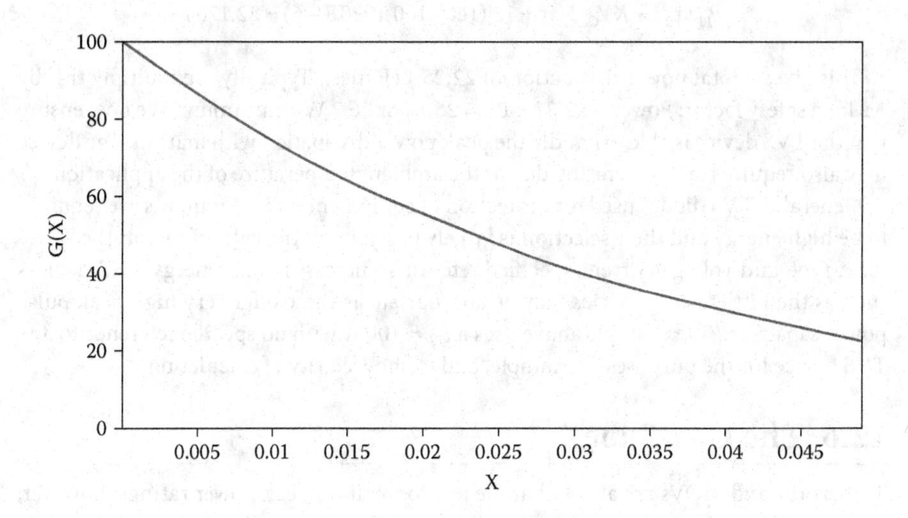

FIGURE 12.9 Exponential pulse, tau based on 50% current, $K = 1.4$.

If the specification for the pulse waveform is equivalent to Figure 12.10, where P_{PP} @10/1000 μs is valid, we may calculate the total power rating as follows:

Using $V_{CL} = 160$ V, $I_{PP} = 100$ A based on the example above,

- Energy of the ramp

$$P_{PP}(\tau_1) = KV_{CL}I_{PP}\tau = 0.5(160)(100)(10E-6) = 0.08J$$

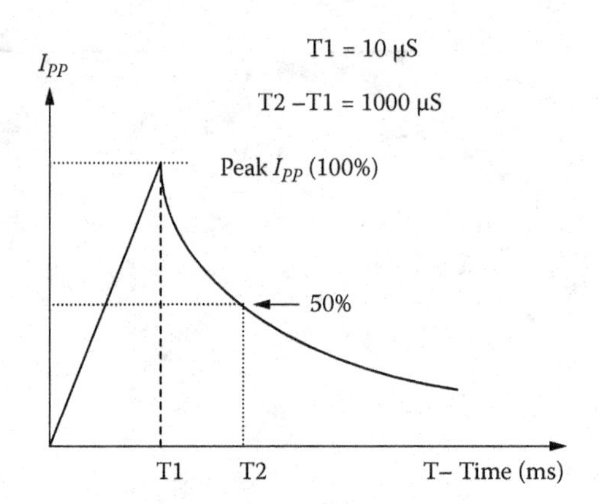

FIGURE 12.10 A ramp and exponential pulse solved together.

- Energy of the damped exponential

$$P_{PP}(\tau_2) = KV_{CL}I_{PP}\tau = 1.4(160)(100)(990E-6) = 22.176\,J$$

This gives a total power dissipation of 22.25 J (Joules). Typically, we multiply this by 1.2 for a safety factor: Power = 22.25 x 1.2 ≈ 26.7 J or 26.7 W-s minimum. We now ensure that the TVS device is able to handle the peak power dissipation with margin. The device may also require further derating due to the ambient temperature of the application.

Generally, TVS diodes used for protection of equipment on power inputs are required to be high energy and their selection is largely based upon the level of threat, the waveform type and voltage/current specified, etc. In some cases, high energy TVS devices such as the Littlefuse AK series may be appropriate as these offer very high peak pulse power capability. The example above uses a $I_{PP} = 100\,A$ with no specific reference to any TVS device for the purposes of example, and to show clarity of calculation.

12.6 The Gas Tube

Transzorbs and MOVs are able to handle low to medium peak power ratings; however, for very high peak power ratings, the use of gas tubes becomes an option. This is particularly so in the case of lightning strikes. These handle 25,000 A over short periods of time. One of the leading companies is Joslyn Electronics, located in Goleta, California, near Santa Barbara. These are best used in DC systems, where the DC voltage will stabilize back to a constant voltage soon after the pulse. The tube will then deionize quickly. In AC systems, because of the continuously changing sinusoidal voltage, the gas tube does not fully deionize. In some cases, the system must be shut down for a short time to rid the tube of the ionization. The application of gas-discharge tubes is specialized, and they are not normally found in EMI filters. As such, they are beyond the scope of this book.

13 What Will Compromise the Filter?

The EMI filter may pass EMC testing using the specified test method and still fail to work as designed for many reasons, the most prevalent of which are discussed here. As mentioned in Chapter 7, section 4, some filters are compromised by ground faults. The case discussed was a balanced filter with the bottom half inadvertently connected to ground, which reduced the loss by 50%. This was caused by lack of communication between two different engineering groups, resulting in double the weight, size, and cost, for half the performance. There are many more candidates for failure, and some of these are discussed here.

13.1 Specifications—Testing

Filters are usually designed to pass the more prominent 220-A test with 50 ohms load and source. An EMI filter was designed for the 220-A specification, but the new customer wanted to test using the current-injection method described by the military standard MIL-STD-461, and the filter failed. This test setup requires two 10-μF capacitors to ground on both the supply and return power leads. The filter was unbalanced, with all the components on the supply, or hot, side. The test setup was detecting common-mode noise, and even though common mode was added, the filter failed the tests. The filter had to be redesigned, and there was little room left. Ultimately, they convinced their customer to change the testing specification.

13.2 Power Supplies—Either as Source or Load

Power supplies are sometimes difficult to deal with either as the source of supply or as the load. As the source, most designs are inductive output, not at the DC output, but they become more inductive as the frequency increases. From 0 to 10 or 50 Hz, depending on the supply, the output impedance is in the milliohm range, and it starts climbing after this and looks inductive. If the EMI filter following the DC supply is capacitive at the input, the inductance of the supply and the input filter capacitor can ring. However, if the EMI filter following the DC supply is inductive input, the stored energy of the inductor has been known to damage the supply on turn-off. The best arrangement for this application is to make the first stage of the EMI filter a T. The inductor facing the supply is half the value of the inductor in the L filter. A back-biased diode across the filter input shunts the voltage to ground. The input filter inductor, including the rest of the inductors, sees reverse voltage; therefore, the line-side inductor is now negative. This turns on the diode and shunts all the voltage stored in the filter capacitor and discharges the energy stored in the inductors. In addition, it also blocks the filter voltage from destroying any part of the source supply through overvoltage.

If the EMI filter is to feed a power supply, a low-output impedance is required. This means the preceding solution with the T cannot be the only filter component. The solution requires an L stage with a quality capacitor, preferably the feed-through type, which is used for the output element. In many cases, the input to the power supply will need a DC link, or hold-up capacitor, and this will form part of the output L stage and provide a low impedance to the power supply. The central inductor can be the total of the T inductor plus the L inductor. In other words, if the calculated value of the inductor was L, the source inductor would be L/2, and the central value would be L + L/2 or 3L/2. This is assuming that the input in both cases is DC. In Figure 13.1, the central inductor is three times the line-side inductor because the central inductor is really two inductors. The T has two inductors, each equal to half the value of the L inductor. So the central design value would call for 1.5 times the calculated value.

What about AC power supplies? Many companies use AC switcher supplies to generate AC at other frequencies such as 50 Hz for Europe and 400 Hz. Some of these generators provide three-phase power outputs. It is important to consider the harmonic content in these AC power sources, and from an EMI filter standpoint, the harmonics can prove to be problematic if both the component selection and robustness is not considered. For example, a 400-Hz supply might have a strong 2,400-Hz component where the filter incurs a resonant rise close to this. The filter components may overheat with little or no load current, as the filter was not designed to handle the 2,400 Hz.

13.3 9- and 15-Phase Autotransformers

Autotransformers can cause problems by themselves if the EMI filter requires substantial low-frequency loss. This occurs when the ratio of the total filter inductance even approaches as little as 2% of the primary inductance and forms a voltage divider, so the output voltage is below requirement. Again, the customer was either not aware that the

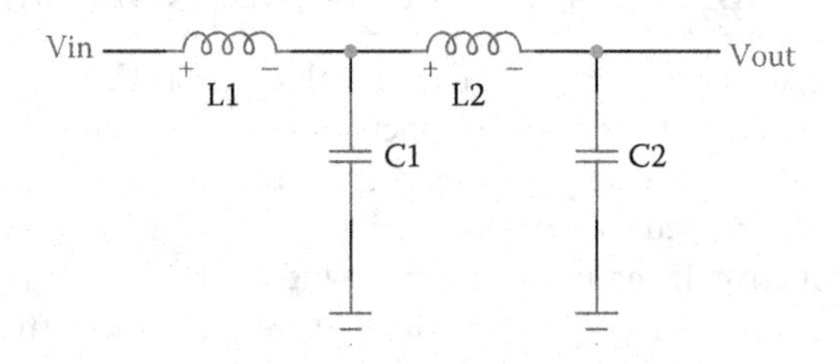

FIGURE 13.1 The input 'T' with the output 'L'.

following device was an autotransformer or was not aware of the problem. The 9- and 15-phase types often start with an autotransformer yielding low primary inductance. The cure for this is to add a filter stage or two to reduce the required total inductance of the filter. For example, in going from a single stage to two stages, the values of both the total inductance and capacitance drop drastically. If this was a single L section with 12 dB per octave, it would change to 24 dB per octave, reducing all the component values greatly. In addition, raising the capacitance and allowing lower total inductance in series across the line may be required. However, this may also play havoc with leakage current specifications.

13.4 Neutral Wire Not Part of the Common-Mode Inductor

In three-phase systems, the currents are always unbalanced. The neutral wire carries this unbalanced current and, therefore, must be part of the common-mode choke. This means that there are four equal windings on the common-mode core. Without this, the ferrite core or nanocrystalline core is driven into saturation because the difference current is not present to balance the core. In this mode, the ferrite core is heated and generates noise that masks the actual noise of the load. Therefore, the noise is sometimes worse with the filter. The same is true in single-phase systems. Both the hot and return legs are wound on the common-mode core. But in many cases there are grounds ahead of the core, compromising equal current flow through this core. Now the core is unbalanced and in saturation, again generating more noise than the load. If this condition exists and cannot be avoided, remove the ferrite core. It will create more problems than it can cure.

13.5 Two or More Filters in Cascade—the Unknown Capacitor

This happens when more than one filter type follows another, as in Figure 13.2. For example, in a secure room or screen room, a large three-phase filter powers the entire room—possibly 500 A per phase. These will be four filter inserts and are enclosed in a larger cabinet on the outside wall. The line-to-line input voltage is 208 and the output voltage is 120 to ground, and these taps are located in the screen room. Note that there are other feeds or taps at the three-phase output (legs in parallel). Each power output leg and neutral may have a filter and possibly other filters in parallel, and each could be rated for 25 A. The one depicted in Figure 13.2 could be fed to cabinets in the room to power a full 19-in. rack of equipment. We should not forget that the individual equipment mounted in the rack is also filtered. This situation places three filters in tandem, or in cascade; furthermore, there could be other filters in parallel. If the power line filter insert feeding the rack required 100 dB at 10 kHz, this may have forced the designer to cut well below the normal recommended cutoff frequency. This would be done to get the proper attenuation or insertion loss for reasonable cost, size, and weight. The filter in the bottom of the rack may be a single π filter and be paralleled by other π filters in the rack.

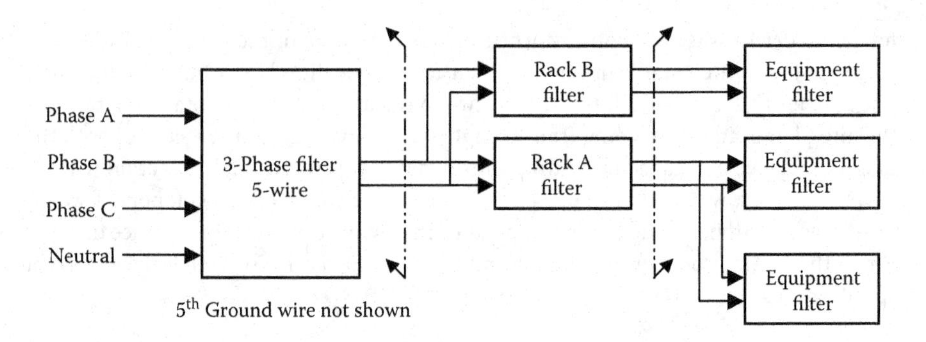

FIGURE 13.2 Three filters in tandem.

These filters detune each other, especially if any or all have higher circuit Q values of 2 or better. The higher circuit Q increases the potential of these filters to oscillate, and this would move the cutoff frequency farther into the passband. The latter problem has been known to reduce the line voltage to a point where the rack equipment failed to work.

If any of the filters were designed incorrectly initially, this would accentuate the problem of cascading. Moreover, it would be worse still if there were multiple double feeds with other filters in cascade connected to the same power line filter. The cascaded capacitors would total in number and cause higher line and harmonic currents that would add heat to the filters. This would increase the number of resonant rises within the filter chain, along with the addition of notches in the frequency magnitude responses, otherwise known as resonant drops. The problems that might very well exist with such an architecture, from both a system and individual filter perspective, should now be obvious.

13.6 Poor Filter Grounding

A properly designed filter may appear to pass well in the EMI test laboratory or at the EMI filter design house. The reader may have been in these laboratories and have seen all the grounding techniques that are necessary to test the filter and system equipment (Figure 13.3). The test bench is covered with a sheet of copper that is well grounded. The equipment, or filter under test is often C-clamped tightly to the copper sheet. Most filters are designed to be mounted directly to a ground plate through input feed-through studs or through the connectors. The filter is mounted through chassis holes with EMI gaskets used on both sides of the chassis.

If this ground is not provided, the filter fails to live up to its dB rating. It may very well be suggested that the filter design is no good even though test results demonstrated that the filter checkout was successful.

The filter is tightened down to ground with the proper nuts and washers to a specified torque. The gaskets give thousands of ground points with this technique. This also carries the chassis ground plane through the cutout holes in the chassis. Without a good ground, the filter's feed-through capacitors and other components to ground cannot work as in Figure 13.3. The Transzorbs or MOVs, including the Y caps that are tied to

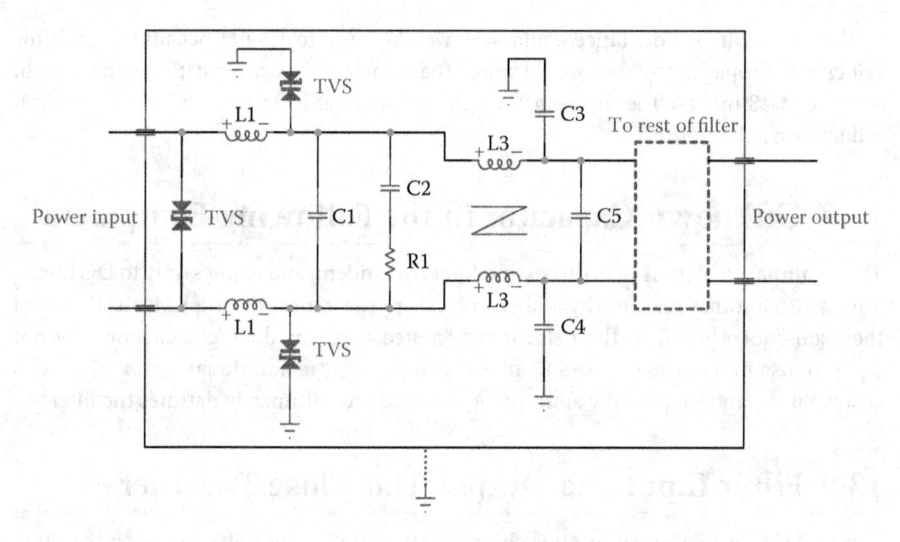

FIGURE 13.3 EMI filter lacking the proper grounding.

ground, cannot function correctly to give the required performance. It is easy to see that the filter in Figure 13.3 does not function properly if the filter case is not grounded. The two load-side feed-throughs would be out of the circuit, along with the two line-side Transzorbs to ground. The two similar capacitors on the input are noncapacitive input terminals or connectors.

13.7 "Floating" Filter

This follows from the topic in section 13.6, where the filter was designed according to the method just described. This filter was to be mounted to ground, and the users complained that the filter was not functioning. Sure enough, the filter was mounted or hung in air through a plastic hanger. A 6-in. green wire (normal hookup wire and not Litz) ran from one lug on the filter through a cable harness to a ground. This was not even Litz wire, which could carry most of the upper frequencies. It was common hookup wire, which acts as a good antenna and was therefore radiating an *H* field that couples into the surrounding wires in the same cable form. Due to cost and schedule, the customer was never able to make the necessary changes. This setup was designed by a mechanical engineer who knew nothing about EMI. They were way behind schedule with their customer pressing them to ship, and could not find time to make the changes necessary. The green wire mentioned here has the following properties:

1. A skin effect adds to the AC, or RF resistance, making the ground more resistive.
2. A slow velocity of propagation, making the apparent length about eight times longer, adds inductance to the lead, further impeding the RF current.
3. From 1 and 2, a series RL circuit is formed.
4. Whatever the RF current magnitude in the lead is, it will radiate. The *H* field will couple into cables and any adjacent conductors in the harness.

The green wire's inductance could very well be close to 1.5 μH because of the slow velocity of propagation. This would make the 6 inches almost eight times the length, or close to 48 inches. The full length would equate to at least a meter, and the typical inductance is 1.5 μH/m.

13.8 Unknown Capacitor in the Following Equipment

This is similar to section 13.5, discussing filters in tandem, and applies only to DC filters. This is also another reason why a filter may be appear to pass during EMC testing, and then subsequently fail in the field. It is accepted that this double capacitor may not compromise every situation in DC filtering; however, the additional component adds cost, volume, and weight. The addition of this capacitor ultimately detunes the filter.

13.9 Filter Input and Output Too Close Together

Many EMI filters designed in-house have the input and output filter ports on the same face of the enclosure, and in close proximity. Sometimes this is done through a need to have one input/output connector, or the customer wiring harness will not allow for a different route for a wire harness. In high-current applications, this is a recipe for failure due to the proximity of both input and output and associated *H* field coupling. In certain applications where, for the purposes of example, a motor controller enclosure is required to have the DC input power and motor phase outputs on the same face, unless the internal architecture of the enclosure provides *H*-field screening, the EMI filter is almost redundant. Again, this is a situation where an enclosure is mechanically optimized without any consideration for EMI needs.

Suffice to say, failure of MIL-STD-461 (CE-101, CE-102, RE-102), for example, would almost certainly demand a mechanical packaging design change and not an adjustment to the EMI filter-insertion loss! With poor mechanical design, forcing all wires through one or two close-proximity apertures will drive a poor electrical layout for component placement and separation. This means that the filter components that make up the input and output sections of the filter are too closely spaced within the filter. Therefore, a filter designed for 40 dB may only provide 26 dB of loss. It is easier for the RF to radiate from input to output or vice versa. Sometimes the input and output are in the same connector and the input and output wires are in the same cable. For example, if the filter is a π structure, the capacitor has very low impedance at the trouble frequencies, which increases the current. If the return is via chassis ground, a large current flows in the wire, which then radiates; the *H* field will couple into all the wires in the cable.

For optimal component placement (Figure 13.4) and filter performance, the best solution is a long filter body with the input power and output feeds on opposite ends. This solution drives a placement approach that allows the filter components to flow from left to right, just as they would appear on a circuit, thereby encouraging a balanced symmetrical placement. In this way, the unwanted energy is dissipated and reflected back to its source as the signals travel through the filter sections toward the opposite end. Figure 13.5 shows an input connector with threads so that EMI gaskets can be used on both sides of the enclosure material. The internal gasket should be the full width and height, and the outside gasket should be the same diameter as the mounting washer. The mounting nut follows this. The output terminals and feed-through capacitors are on the far end. In cases where the filter must have both inputs and outputs on the same face, it would be necessary to place a shield between the two halves of the filter to reduce coupling. In this scenario, it is very important not to allow the harnesses feeding the filter to carry both feeds.

There are two ways to divide a filter enclosure, and the choice is determined by the component sizes, magnetic cores, etc. The first is to divide the width by two and install a shield that runs well to the other end of the filter, but leaving a small opening for the components to loop back up the other side (Figure 13.6). This makes the shield the same height as the enclosure to ensure separation. The input or output section is now half the effective volume. The input studs and input filter components should be installed on one side of the filter. The rest of the filter components should be placed so that they flow toward the rear and then turn the corner and head toward the front again. If some components are larger than the compartments formed by the shield, the second

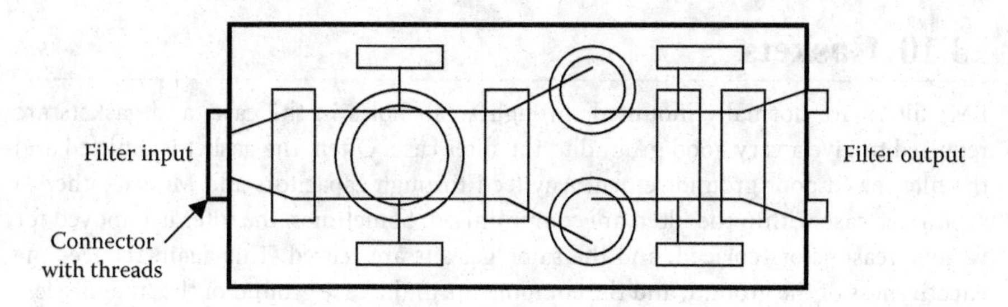

FIGURE 13.4 Filter component placement – symmetrical flow.

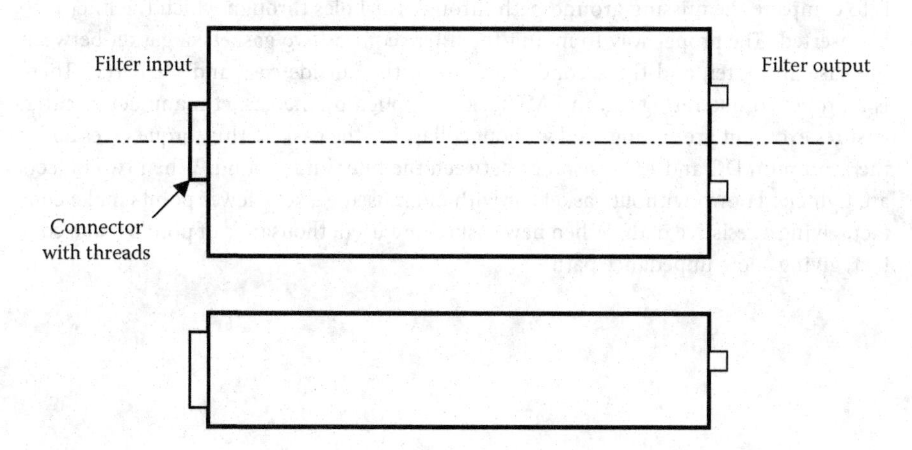

FIGURE 13.5 Filter assembly with input and output connectors.

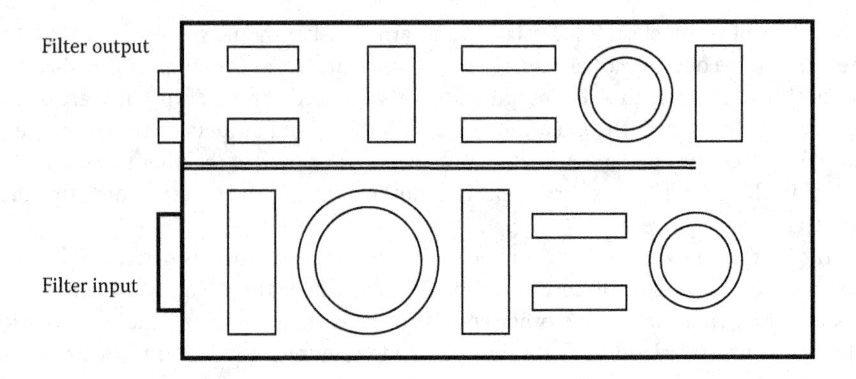

FIGURE 13.6 Filter component placement – wrap-round flow.

method of using the shield to split the 2-in. height can be used. Continue with the same technique.

If the components still do not fit, it is best to change the layout so that the filter can run from end to end without doubling back, as seen in Figure 13.4.

13.10 Gaskets

EMI filters are normally mounted through small holes in the case, and gaskets are required to give a very good ground to the filter case. Often, the gasket is omitted and the filter has a poor ground return. Any feed-through capacitors and MOVs, either to ground or case within the filter, are compromised. Sometimes, the filter is removed for various reasons or replaced, and the same gaskets are reused. This again reduces the effectiveness of the ground, and the components to the case ground of the filter are less effective. On the other hand, it is better to reuse the gasket than to be without any gasket. A gasket presents thousands of ground contacts between the filter case and container, whereas a filter without a gasket may have 5 to 10 contacts. Another reason to use gaskets is to complete the missing ground path through the holes through which the filter is to be inserted. The proper way to mount the filter requires two gaskets: a gasket between the case and filter, and the second set between the outside case and washer(s). There is a proper torque rating for an EMI gasket supplied by the gasket manufacturer that ensures excellent grounding, and without collapsing the gasket. This torque gives about the minimum DC and AC resistance between the filter and ground. When two objects are tightened down without gaskets or with old reused gaskets, fewer points make contact, giving a resistive path. When new gaskets are used, thousands of points make contact, giving a low-impedance path.

14

Waves as Noise Sources

The waves (transient) more commonly encountered within systems are discussed in this chapter. These waves are not as pure as drawn here; some parasitic oscillations from the transformer and other components will be superimposed on the waveforms. In real-world observation, these waveforms have rise and fall times due to parasitic effects; however, they are shown in the text as pure step functions that are impossible to achieve. Each voltage, or current, waveform is derived from the equivalent Fourier equations.

14.1 Spike

Voltage and current spikes are one of the common waveforms or noise sources encountered in electronic circuits, as shown in Figure 14.1. Voltage spikes are short-duration impulses in excess of the normal voltage. Although their duration is often very short, voltage spikes may exceed the normal voltage of a circuit tenfold or more. The cause of low-energy spikes is mostly attributed to switching circuits that have inductance, either parasitic or within the load. In the case of a high-voltage three-phase bridge, for example, rapid turn-off of the high-side power switch may result in a high dv/dt due to the commutation inductance. This is a voltage spike and may cause power device avalanche if the device rating is exceeded. Lightning strikes can also create voltage spikes, even when the discharge event may have occurred a mile away. The transient voltage is transmitted through utility lines and can be seen at the input of electrical equipment.

In the case of a current spike, for example, they are seen when double-ended, or Royer, circuits are used, most often when one switch is turning off while the other is turning on. During the delay of transformer core saturation at the end of switch-A on-time and the turn-on of switch-B, the off-going switch operates with high drain voltage, and this will create a short but potentially high current spike, which may damage the switch. The two currents add, and as far as the switchers are concerned, the current nearly doubles. Other elements such as diodes also add to this current, and often the total spike current is many times the average switcher current.

This spike occurs twice per cycle of the switcher frequency; therefore the frequency of the spike is double the switcher frequency. Without some form of holdup correction in the switcher circuit in a DC system, this high spike current can result in excessive DC

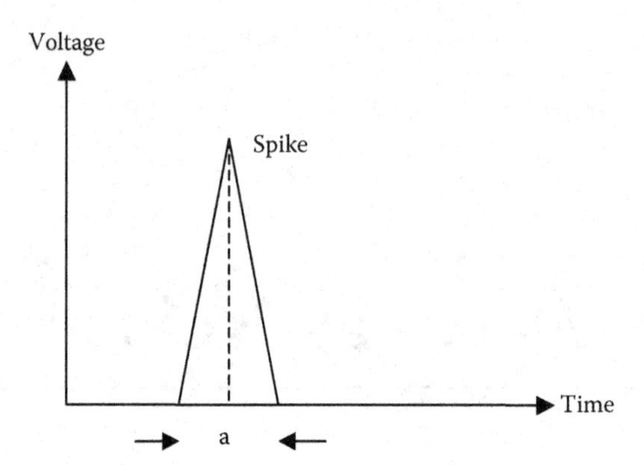

FIGURE 14.1 Spike.

voltage droop. This impairs switcher operation, and most of this spike energy is passed back to the EMI filter. Adding additional DC link capacitance will help to reduce the effects of this spike, and the EMI filter will not see the full effects.

With an off-line regulator, the storage capacitor is often very large, in the 500–2000-μF range or more. If the spike frequency is, for example, 140 kHz—twice the 70-kHz switcher frequency—this large storage capacitor may be well above its self-resonant frequency (SRF) and be either resistive or inductive. If this holdup circuit is not implemented, then a good quality capacitor should be placed across the DC link as close to the switcher as possible.

A simple method to calculate the line-to-line capacitor is as follows: Estimate the lowest working voltage of the DC link and divide this by twice the peak switching converter current. This equates to the working impedance of the switching converter. The capacitive reactance of this DC link capacitor should be 10% of this working impedance at the spike frequency, or 140 kHz in the case of the example. For practical implementation purposes, round the calculated value for capacitance up to the next higher standard value. This is a pulse and, as such, the capacitor chosen must handle the higher pulse currents; therefore foil, not metallized film, should work with the proper derating. The SRF of this capacitor must be greater than 10 times the switcher frequency, or 10 × 140 kHz = 1.4 MHz.

Ceramic capacitors also work very well here; however, the leads must be short because a capacitor with the same lead length will not give the higher SRF required. The equation of the spike is as follows:

$$\frac{Ea}{T}+\frac{Ea}{T}\sum_{N=2,4,6,8}^{\infty}\left|\frac{\sin(X)}{X}\right|^{2}\cos\left(\frac{2\pi Nt}{T}\right) \tag{14.1}$$

where a is the pulse width and

$$X=\frac{2\pi Na}{T}$$

14.2 Pulse

The pulse is similar to the quasi-square, but with pulses in the same direction as for the spike shown in Figure 14.1. This is similar to the Royer, where the dwell time for both halves is not on for the full half-period. Therefore, this generates pulses of current twice per period, once for each half, and is again at twice the switcher frequency. The design and considerations are the same as for the spike. The equation of the pulse is

$$\frac{Ea}{T}+\frac{2E}{\pi}\sum_{N=2,4,6,8}^{\infty}\frac{1}{N}\cos(\pi Nt)\sin\left(\frac{\pi Na}{T}\right) \tag{14.2}$$

where a is the pulse width (Figure 14.2).

The quasi-square is the wave applied to the gate, or control, of the switch, and the reciprocal is applied to the opposite control, as shown in Figure 14.3. The pulse width is a, and the dwell angle is one-half the angle between the pulses. The dwell angle occurs before and after each pulse, or four dwells per period. This turns on the opposite device every half-cycle, and each is turned on for less than a half-cycle. The output is the current pulse at twice the quasi-square frequency. The pulse is filtered by the technique described in section 14.1 for the spike.

This quasi-square wave and its output wave, the pulse, are not real-world waves. Both lack the rise and fall time that are evident in a real-world application. This is due to parasitic capacitance and inductance within the circuit. All devices such as circuit boards, the natural capacitance of wiring, and the input capacitance of the switch have properties that require time to discharge or change. This results in ramps in the wave shape, changing the so-called square, quasi-square, and pulse into trapezoids that show the rise and fall time. If we factored in the parasitic inductances, we would also see high-frequency ring. The equation of the quasi-square wave is

$$\frac{4E}{\pi}\sum_{N=1,3,5,7}^{\infty}\frac{1}{N}\sin(n\omega t)\cos(N\phi) \tag{14.3}$$

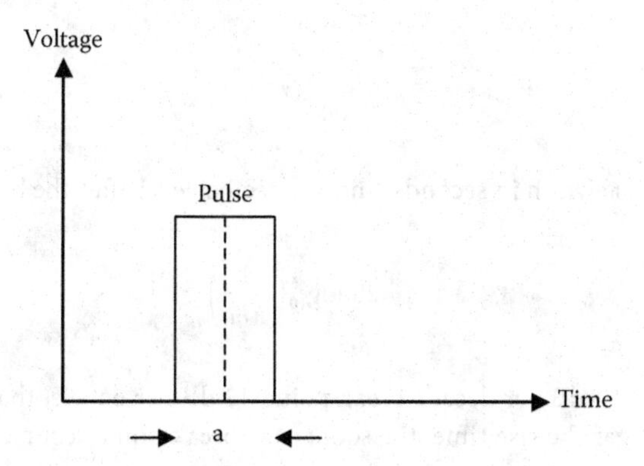

FIGURE 14.2 Pulse.

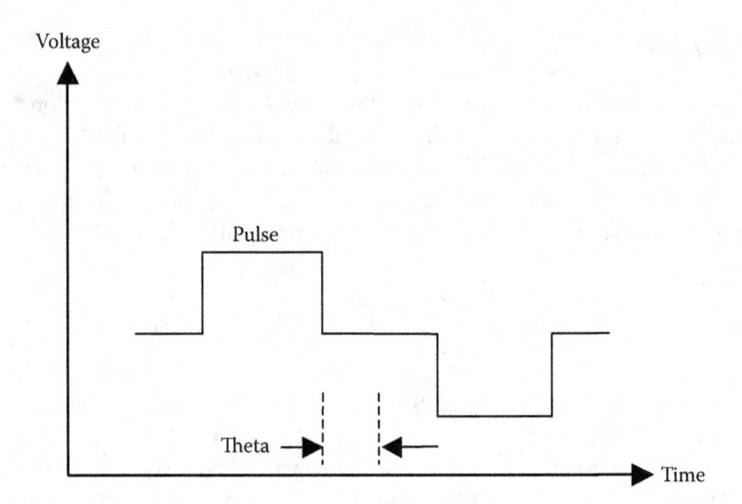

FIGURE 14.3 Quasi-square.

14.4 Power Spectrum—dB μA/MHz

The question arises: "How much power exists within the various waveforms discussed in the earlier sections?" These form envelopes that provide the peak power, which varies with the amplitude of the current, I_p. If the current pulse width is a and T is the period, the equation for the pulse power dB/MHz is as follows in Figure 14.4.

$$P = 20\log_{10}\left(\frac{2Ea \times 10^6}{T}\right) \tag{14.4}$$

where E is the amplitude. This gives a flat line across the frequency spectrum to the 20 dB per decade, or 6 dB per octave, break point. This point starts at the frequency

$$\frac{1}{\neq a}\text{Hz} \tag{14.5}$$

where a is the pulse width in seconds, and the decibel level after the break point is

$$P = 20\log_{10}\left(\frac{1}{\pi Fa}\right) \tag{14.6}$$

There is also a 40-dB per decade break point, 12 dB per octave, that depends on the rise time. The larger the rise time, the sooner the break point occurs. The point here is

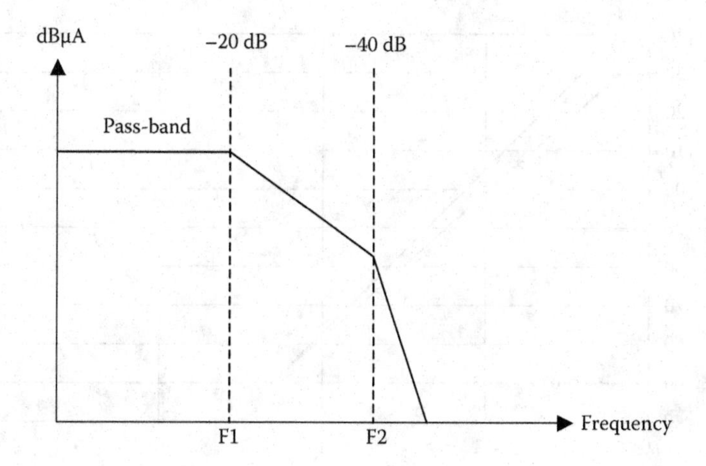

FIGURE 14.4 Power spectrum.

that EMI energy can be greatly reduced by using this principle. The efficiency of the power supply decreases, but the EMI energy is less. This is usually a better trade-off for smaller and lighter units and should reduce the total cost. If the rise time is 10% or better, this becomes, in essence, an L filter added to the existing EMI filter at frequencies above the 40-dB break point.

$$\frac{1}{\pi\tau}\text{Hz} \tag{14.7}$$

The added loss after the 40 dB per decade break point is

$$40\log_{10}\left(\frac{1}{\pi^2 F^2 \tau}\right) \tag{14.8}$$

where is F in hertz, and **τ** is in seconds.

14.5 MIL-STD-461 Curve

This specification is in dB **µ**A/MHz rather than insertion loss. See Figure 14.5. It is rather difficult to convert from this to insertion loss. If the impedances were equal, the conversion would be the obvious 120 dB. Insertion loss is stated in terms of how much loss is required, whereas dB **µ**A/MHz, or dB **µ**V/MHz, is in terms of how much noise is allowed. There is a 13-dB conversion factor due to changing from 50 ohms to

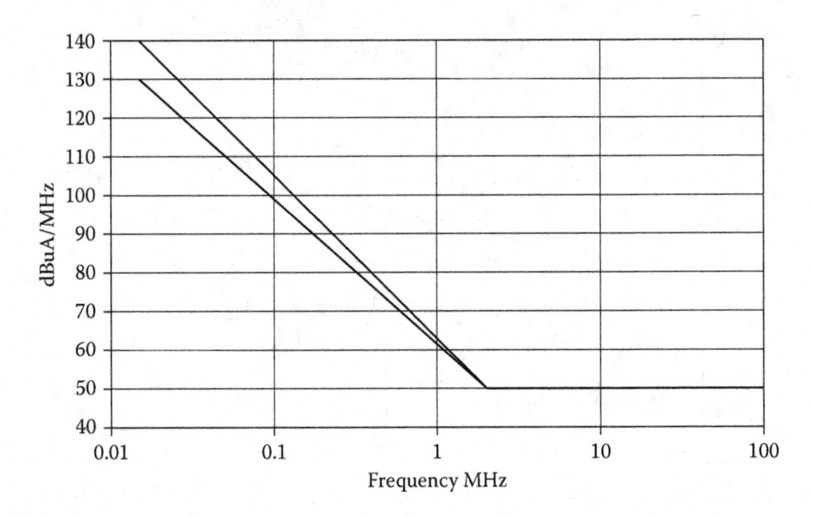

FIGURE 14.5 MIL-STD-461 curve.

a probe estimated to be 1 ohm. This subtracts from the 120, giving 107 dB, and then the probe correction factor is needed. This is a function of the probe and the frequency bandwidth. This last correction is added back to the 107 dB. Usually, this method comes with the probe. The top curve in Figure 14.5 is for the navy, and the bottom is for the army.

15

Initial Filter Design Requirements

With all applications, a filter design must start with a set of requirements, and the performance requirements of the filter are critical to design success. When considering a new filter, there are many constraints that will dictate the design drivers, and these are as follows:

1. Equipment application
2. EMC performance requirements
    a. CISPR, DO-160, MIL-STD-461, other
3. Mechanical constraints
    a. Form factor
    b. Weight
4. Business constraints
    a. Cost
    b. Use of standard/COTS parts
    c. Schedule
5. Input power source
    a. AC single or three phase
    b. DC single ended or floating (differential)
    c. Inrush requirements
    d. Lightning requirements
    e. Power-interrupt requirements
6. Output power load requirements
7. Switching frequency in the case of switching converters
8. Differential-mode design goals
    a. May be defined through analysis
9. Common-mode design goals
    a. May be approximated but often difficult to define
    b. Robust design approach within switching circuits will alleviate common-mode effects

Once we have a robust set of both needs and requirements, the filter design may start with top-level design risk mitigation. These are as follows:

- Differential-mode goals
  Develop differential-mode loss approximation
- Common-mode goals
  Estimate of common-mode load impedance
- Consider methods for reducing the size of the inductor related to inductor current
- Define filter structure
- Simulate performance
- Packaging

15.1 Differential-Mode Design Goals

Ultimately, the EMI filter must pass EMC qualification, and the differential-mode loss requirement is the first design challenge. To be able to achieve this, it is necessary to make a close approximation of the dB loss by analysis. This is a relatively straightforward procedure and will use the magnitude of the fundamental harmonic as a limiting factor such that the loss at this frequency should be equivalent to this magnitude along with a margin for error, etc. The process of analysis is simplistic and may use a close approximation of the current signature for the switching converter at a frequency (f_{sw}). If the peak current amplitude definition is accurate along with the shape of the current, then by performing a fast Fourier transform (FFT) sweep, the harmonic content may be captured. The dB/frequency limit asymptote, e.g., DO-160, etc., may then be overlaid onto the FFT spectrum, and from inspection, the filter design requirements for differential-mode loss may be defined.

Of course, there are other factors that might also effect the harmonic composition, and these are related to circuit factors and parasitic elements that may or may not be part of the model construct used for analysis. The quality of the model will determine the accuracy of the data.

The idea is to make the differential-mode filter transparent to the line. This is ideal for both DC and AC systems and easy to accomplish at DC, 50 Hz, and 60 Hz. However, it is a demanding task at 400 Hz in any system requiring substantial loss at low kilohertz frequencies. The requirement stated really means that the load impedance is transferred to the input of the filter at the line frequency and most of its lower harmonics. The harmonic content depends on the quality of the line and the load. The higher the line impedance, the more effect the odd-order harmonic content has in distorting the sine wave voltage shape. These harmonics are odd harmonics. The filter cutoff frequency to accomplish this goal should be above the 15th harmonic because the level of any harmonic above the 15th, even for the poorest quality line and load, is insignificant. This is why the rule is set at this harmonic of the line frequency. This is easy to do for both 50 and 60 Hz, where the cutoff frequency will be well above this goal; however, the problem is 400 Hz. In fact, the 15th harmonic is not high enough for 400 Hz because of

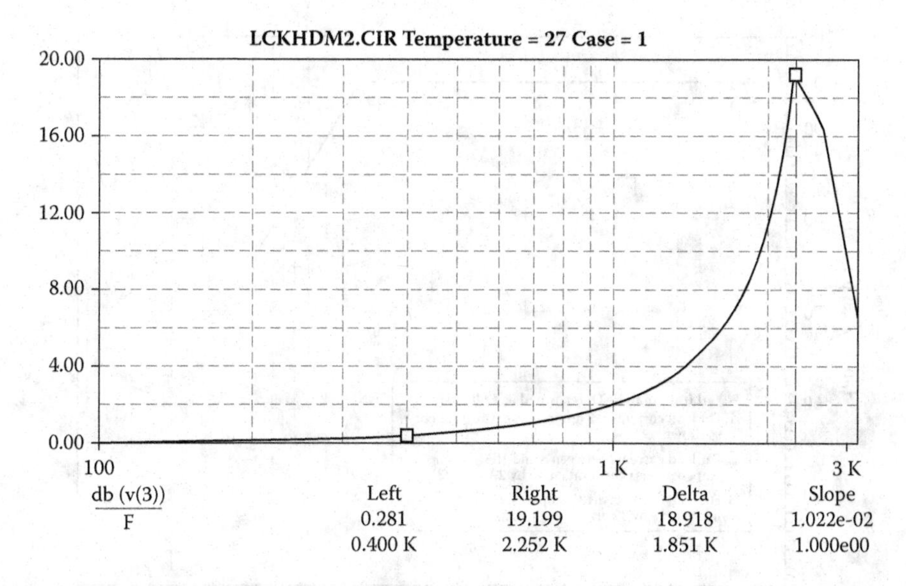

FIGURE 15.1 Resonant rise at 2252 Hz affects 400 Hz.

the resonant voltage rise at 400 Hz. In a perfect world, it would be prudent to place the cutoff as high as possible to 8 kHz. However, this requires two or more stages to meet the insertion loss needed. In the past, the cutoff was formulated at the 10th harmonic, but this did not take the 400 Hz into account.

In Figure 15.1, the cutoff frequency is well above 4000 Hz but has a serious resonant rise at 2252 Hz. This gives a 0.281-dB gain at 400 Hz, which translates to a 3.3% voltage rise. This is calculated for ideal conditions, and the true rise is much more than 3.3% in operation. Note that this condition is not a problem at 50 or 60 Hz, as the dB gain is zero. Actually, the response in Figure 15.1 is based upon a 60-Hz filter, so there would be no problems using this, as there is insufficient power at 60 Hz.

With MIL-STD-461, 400 Hz is often an issue, particularly so when we consider the needs of CE-101 (Figure 15.2) and in maintaining sufficient loss: 95 dBμA between 30 Hz and 2.6 kHz, and 76 dBμA or −10.5dBV at 10 kHz. In addition, CE-102 (Figure 15.3) has a relatively tough requirement where conducted emissions starts at 10 kHz. With no relaxation, the loss at 10 kHz is 94 dBμV, or −26 dBV. Achieving this is sometimes difficult to do as the number of stages required is often greater than two.

15.2 Differential-Mode Filter Input Impedance

The equation is simply Z_{in} at $N_h = R_{load}$, where Z_{in} is the input impedance of the filter, N_h is the harmonic number and is set equal to 15, and R_{load} is the load impedance at the same frequency. Practically speaking, the load impedance is constant over the frequency range of interest in this discussion.

This design goal is often difficult to reach, especially for the high-current filters, where the required loss is heavy at low frequency. If the cutoff frequency allows the filter to

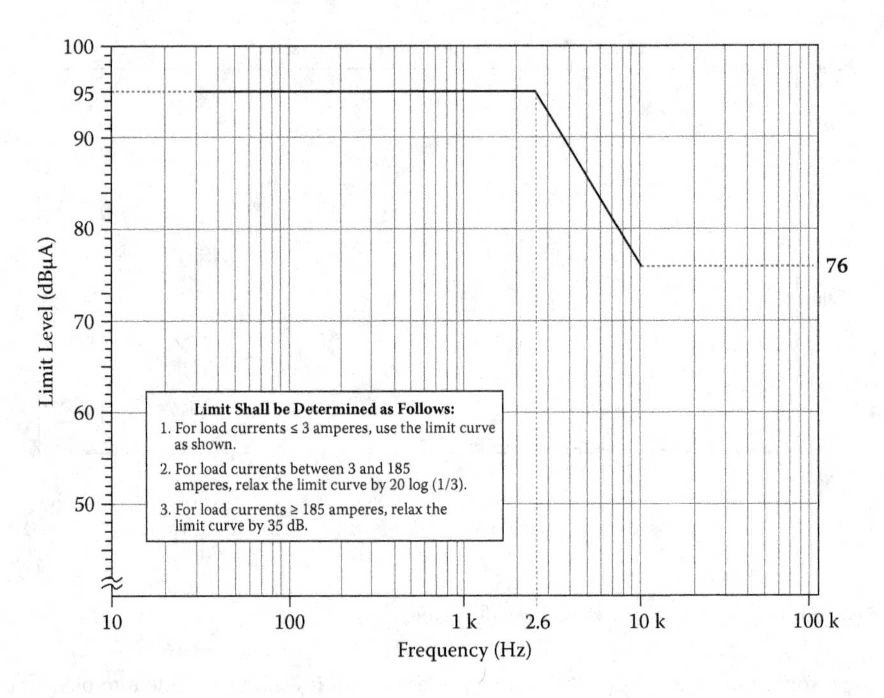

FIGURE 15.2 CE-101 for DC applications.

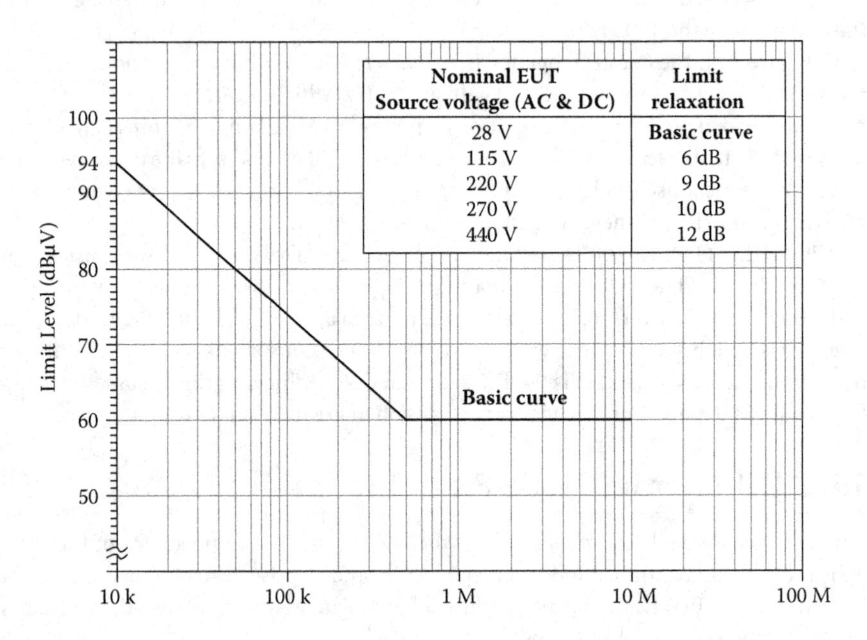

FIGURE 15.3 CE-102 limit (EUT power leads, AC and DC) for all applications.

attenuate the lower harmonics, higher capacitor currents result, which heat the capacitors because of the equivalent series resistance (ESR). The low-frequency cutoff also increases the harmonic currents through the inductors, which increases the temperature of the inductor due to DC resistance (DCR) and higher core losses. Overall, this has the effect of raising the operating temperature of the filters, which are typically used for power line filtering. The low cutoff frequency will also lower the resonant rise frequency and raise the circuit Q. Either one of these facts could result in a difficult-to-tune filter that is subject to high operating temperature.

15.3 Differential-Mode Filter Output Impedance

This section follows the same logic as for the input impedance of the filter. The filter should be transparent to the load. If the input impedance goal is met, the output impedance goal is normally met. Meeting these two goals makes for better filter operation.

Z_o at $N_h = R_s$, where Z_o is the filter output impedance, R_s is the line impedance at the same frequency, and N_h is the harmonic number. The line impedance here is the basic DC resistance of the line. This holds true for most lines to 5 kHz before any rapid increase to higher line impedances is reached. At 10 kHz, the impedance is about 4 ohms on most lines.

15.4 Input and Output Impedance for a DC Filter

Both requirements of the two preceding sections are easily met for a DC system unless the load is a switching converter. Here, the output impedance of the filter must be very low at and above the switcher frequency. This statement rests on the premise that the switching converter has been designed with no front-end correction. Obviously, this is easy to do if both the EMI filter and switching converter are designed by the same person or team. In principal, the output impedance of the filter, $Z_0 \ll R_L@F_{sw}$, where F_{sw} is the switch frequency and the rest of the terms are the same. The same holds true at the 10th harmonic of the switcher frequency. The switcher may not be starved at the fundamental and yet be starved at the 9th or 11th harmonic if the output impedance is slightly inductive or the output capacitor is above its self-resonant frequency (SRF). This is often described as incremental negative resistance. A switching converter is designed to hold its output voltage constant even though the input voltage is not constant. Given a constant load current, the power drawn from the input supply is therefore also constant. If the input voltage increases by some factor, the input current will decrease by this same factor to keep the power level constant.

In incremental terms, a positive incremental change in the input voltage results in a negative incremental change in the input current, causing the converter to look like a negative resistor at its input terminals. The value of this negative resistance depends on the operating point of the converter according to

$$R_N = -\left[\frac{V_{IN}}{I_{IN}}\right] \tag{15.1}$$

where R_N = negative resistance, V_{in} = input voltage, and I_{in} = input current.

If the impedance of the filter is higher, the switching converter is starved. The output impedance of the filter should be on the order of 10% of the load impedance during conduction or the "on" time. This is also a function of the pulse width. The main goal is to make sure that the drop is not excessive so that the switcher can function properly. As an example, F_{sw} = 80 kHz, V_s = 28 V, and I_{on} = 8 A during the on time (peak). Under this condition R_N = 28 V/8 A = −3.5 ohms. To ensure the switcher sees low impedance, $Z_S << Z_L$, we add a capacitor across the input. Having defined the load, or switching converter negative resistance, we are able to define the minimum capacitor impedance necessary to ensure satisfactory operation of the switcher.

$$R_N = -3.5\Omega \rightarrow \frac{R_N}{10} = (j\omega C)^{-1} = 0.35 \quad C = \frac{1}{2\pi F_{SW}(0.35)} = 5.68\,\mu F \tag{15.2}$$

In most cases, the input to the switcher will have sufficient hold-up capacitance added to ensure that, during full-load switching, the DC link voltage droop is within acceptable limits. It is important to note that the switching converter input impedance, Z_I appears as a negative resistance only at low frequencies. At higher frequencies the impedance is influenced by the converter's own internal filter elements and the limited bandwidth of its feedback loop. If we look at the relationship of both source and load impedance, we can write an expression that equates to a voltage divider of the source impedance interacting with the load, or negative resistance. Furthermore, the poles of the source and load will ultimately determine the dynamic behavior of the second-order system, which is equivalent to a quadratic polynomial. If we connect the switcher to source impedance Z_S, the frequency-dependent terms that govern the switcher's performance may be multiplied by the relationship in equation (15.3).

$$\frac{Z_I}{Z_I + Z_S} = \frac{1}{1 + (Z_S / Z_I)} \tag{15.3}$$

If $|Z_S| << |Z_I|$ for all frequencies, we can say that the effects to the operation of the switcher in terms of stability and performance are negligible. To ensure stability, we must ensure that the poles of equation (15.3) lie in the left-hand S-plane. Typically, the EMI filter, or input filter, will have a series inductor and include DCR ohms; therefore, the source impedance equates to $Z_S = sL + R_{dc}$. The load impedance is the capacitor in parallel to the equivalent negative resistance, as follows:

$$Z_L = R_N \| C = \frac{R_N}{1 + sCR_N} \tag{15.4}$$

For a stable system,

$$sL + R_{dc} \ll \frac{R_N}{1 + sCR_N} \tag{15.5}$$

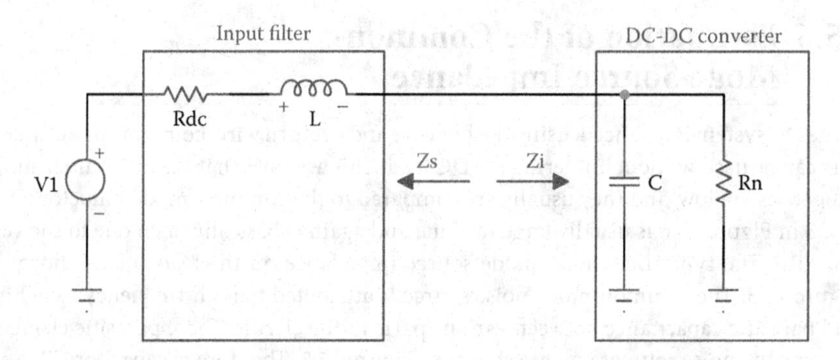

FIGURE 15.4 Source and load impedances.

If R_{dc} is added to reduce Q, then a constraint for a stable system is $R_{dc} << |R_N|$ (see Figure 15.4).

From a practical standpoint, the filter would be stabilized by using a damping dQ RC shunt network, and this is discussed in chapter 19.

15.5 Common-Mode Design Goals

The common mode does not have to meet any of the requirements discussed for the differential mode. The cutoff frequency can be as low as desired, cutting well into power harmonic frequencies. It is often necessary to watch for leakage inductance in the Zorro inductor(s). The fluxes from each coil that do not pass between the two coils, but instead through the high-reluctance air path outside the toroid, are referred to as leakage fluxes. The leakage fluxes that leave the coils and pass mostly outside the toroid must return to their respective coils. However, these fluxes pass back through the coil through part of the toroid. With an increase in frequency, the permeability of the core decreases and the reluctance of the path through the coil increases. As the reluctance of the core increases, a greater percentage of the flux produced from the coil passes outside of the toroid, resulting in greater leakage.

The disadvantage is that the common mode grows to very large sizes as the cutoff frequency is lowered, but there is no lower-frequency limit for bandpass or other reasons. The real limit is set by the current rating and size of the common-mode inductor. These inductors should be designed to have little effect on power factor correction circuits, switchers, or any other load. Again, this assumes little differential-mode inductance within the common-mode inductor. The typical amount is 1% to 2% and contributes toward the differential-mode inductance and, therefore, the differential-mode loss. Common-mode inductors are often in the range of 0.5 μH to 33 mH, and beyond. At 2%, this equates to 660 μH, which is often greater than the differential inductor value in some EMI filter applications. Therefore, the common mode often helps with the differential losses, but the reverse is also true if the circuit is balanced with inductors in both the hot and return lines. Some companies do this on purpose by separating the windings and thereby reducing the coupling. The ferrite toroid, in which the common-mode windings are each distributed over half the core, has this characteristic.

15.6 Estimation of the Common-Mode Source Impedance

If the DC system is balanced, using the hot wire and a return wire, common-mode inductors can be used without hindering the DC load. This assumes that the differential-mode properties are low, and they usually are compared to the common-mode inductor.

C3 in Figure 15.6 is usually large in value and again is basically open due to the very low SRF. The typical common-mode source impedance on the load side is shown in Figure 15.5. The common-mode noise source is attributed to high-frequency switching and parasitic capacitance between various parts of the circuit. These parasitic elements are now the noise sources and are shown in Figure 15.5. The storage capacitor C7 is out of the circuit by now due to a low SRF at the parasitic frequency. The noise source could be the stray capacitance between the switching device and ground; the diodes also have capacitance and also contribute to the common-mode noise, as does the parasitic capacitance in the transformer. The equivalent circuit looks similar to that of Figure 15.6, where the current probe measures the noise. Figure 15.7 shows the four capacitors across the diodes and the output transformer capacitor. If the transformer has a Faraday screen, the final capacitor is even smaller. All of this makes the common-mode circuit impedance much higher than the differential-mode impedance (Figure 15.8).

In addition, the circuit impedance is greater than the current probe until the upper frequencies are reached. If the circuit lacks the transformer, the diode capacitance to ground is still much higher. Figure 15.9 shows that the common-mode design impedance is higher than the differential-mode impedance in most applications. This information allows the following technique. Calculate the cutoff frequency as before, and then determine the inductor and capacitor values with the same equations.

$$L = \frac{R_d}{2\pi F_0} \qquad C = \frac{1}{2\pi F_0 R_d} \tag{15.6}$$

where R_d is the design impedance, which is the same as for the differential mode, and F_0 is the cutoff frequency. If there is a leakage current specification, calculate the capacitor value of the capacitor to ground, or the specification may state the maximum value of capacitance to ground. Divide the needed value of capacitance by the maximum value,

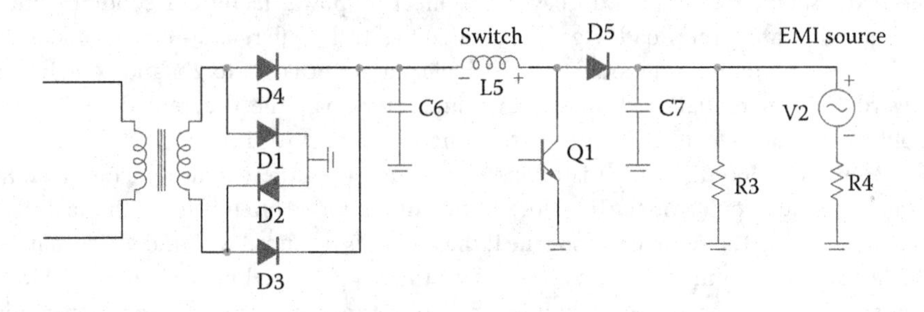

FIGURE 15.5 Load-side common-mode source—high Z.

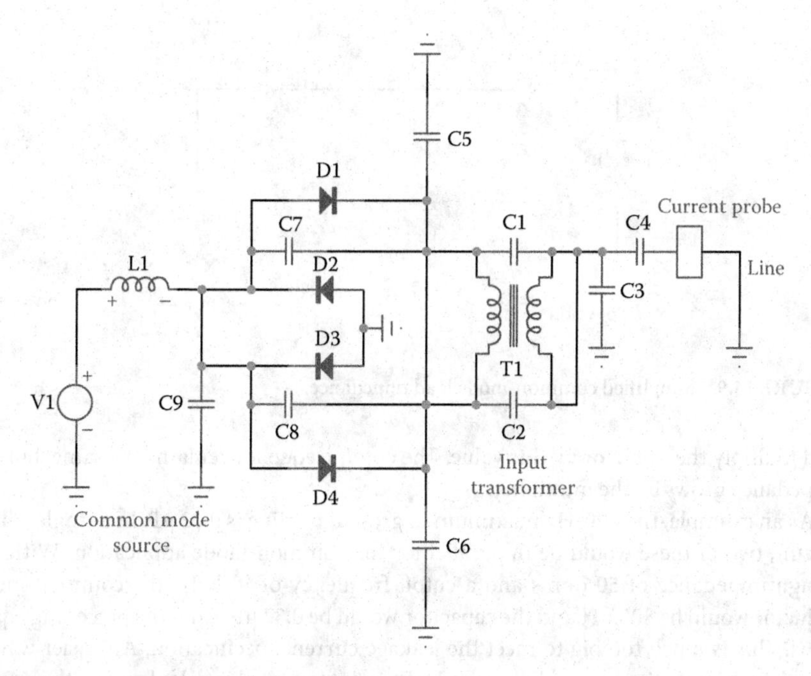

FIGURE 15.6 Common-mode impedance is high.

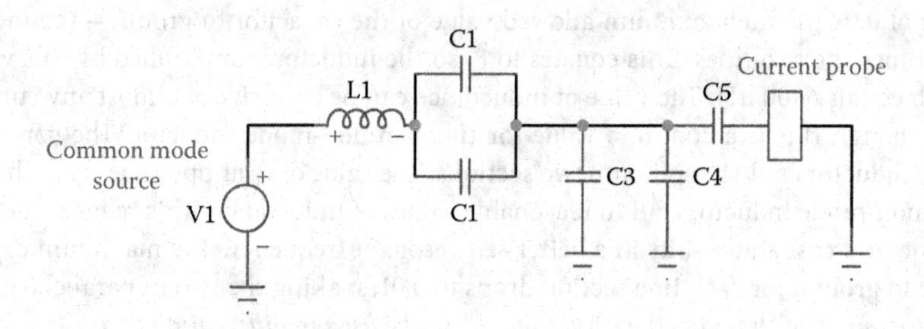

FIGURE 15.7 Simplified circuit of Figure 15.6.

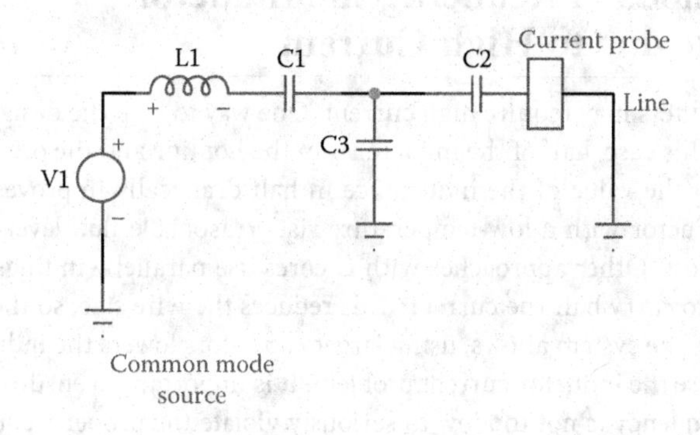

FIGURE 15.8 Switching converter reduced to capacitors.

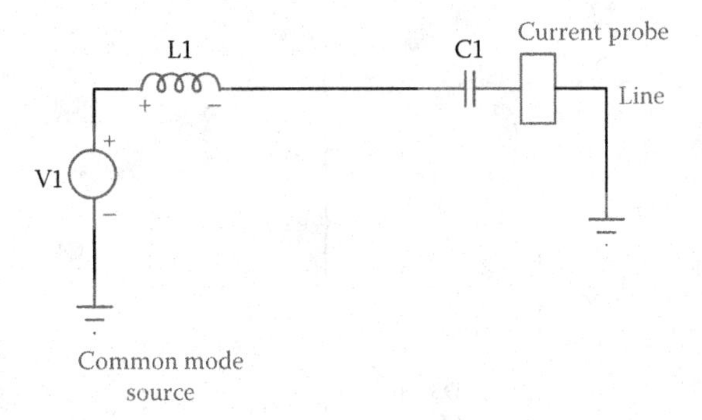

FIGURE 15.9 Simplified common-mode load impedance.

and multiply the inductor by this value. The cutoff frequency remains the same, but the impedance grows by the multiplier.

As an example, the 400-Hz maximum to ground per line is 0.02 μF. In a single-phase circuit, two of these would be in parallel for the common-mode application. With the design impedance of 50 ohms and a cutoff frequency of 10 kHz, the common-mode inductor would be 800 μH and the capacitor would be 0.32 μF using the preceding equations. This is much too big to meet the leakage current specification. An easier way to calculate the capacitor would be to divide the inductance value calculated by the square of the design impedance, or 2500 ohms. Divide this value of capacitor by the maximum value of 0.04 μF, the maximum allowed value of the capacitor to ground—the total of 0.02 times the two lines. This equates to 8, so the inductor is multiplied by this value, which equals 6400 μH. The value of inductance can be excessive at almost any current, even though this is a common value for the common-mode inductor. The common-mode inductor could be split into two sections. The value of F_0 jumps and makes the two common-mode inductors fall to reasonable values of inductance. This reduces the size, weight, and cost and results in a better self-resonant frequency. The maximum capacitance to ground for each line section drops to 0.01, making the two in parallel 0.02 μF. *This technique of changing the values should not be used in differential mode.*

15.7 Methods of Reducing the Inductor Value due to High Current

Some of these filters may require high current. One way to help the design is to balance the circuit. In this case, half of the inductor is in the hot line and the other half is in the return. Cutting the value of the inductance in half drastically improves the ability to design this inductor with a low-temperature rise, reasonable flux levels, and possibly at reasonable cost. Other approaches with C cores use parallel windings. Each arm or side is wound to carry half the current. This reduces the wire size, so the wire is easier to wind. Where the system allows, using larger capacitors lowers the inductance, which also helps to ease the inductor current problem. It is important to ensure that the calculated cutoff frequency is not too low to seriously violate the proper frequency distance from the line frequency. Otherwise a resonant rise will occur at the line frequency.

16

Matrices, Transfer Functions, and Insertion Loss

This book is not a mathematics or matrix course, however the engineer needs to be familiar with the methods used to get to the filter component values in chapters 16 and 17. The *A*-type matrix is both simple and practical for solving many electrical problems. Section 16.2 will look at the matrices of various topologies.

16.1 Synthesis, Modeling, and Analysis

There are several excellent programs that can be used to both define and verify filter values as follows: frequency magnitude loss at any frequency (especially those frequencies out of limits), the proper frequency magnitude slope, and the possible phase shifts over the band of interest. The same is true for modeling the circuit. Some people will spend days using various programs and still not have suitable component values necessary to build the filter. The EMI filter designer needs simple and effective methods of reaching the component values necessary for a baseline design. Thereafter, with an element of defensive design, and adjustment of the filter network during test, the EMC test should prove to be successful.

Let us assume that the equipment has to meet a CISPR specification of 60 dB at 150 kHz: We might therefore assume that the filter requires this level of loss—but *no*, not at all! This is a classic misconception and one that can often lead to failure. In a perfect world, the system needs to be tested for conducted emissions first, and without a filter. In 95% of cases, the filter will need to be designed up-front and probably in parallel with the equipment it is supposed to protect. If the equipment is not able to be evaluated for emissions, then we can use simulation, numerical methods, and trial and error based upon past experience with similar equipment.

If we assumed that a test was run before the filter was designed. The equipment may need 12 dB of additional loss at 150 kHz and a few other points across the spectrum. Note these frequencies and the amount of extra loss needed. If there are a group of outages in one frequency band, log the center frequency and the dB limit outage. A filter requiring 12 dB at 150 kHz is much easier and cheaper to build as compared to 60 dB at 150 kHz. Now add 6 dB of loss at 150 kHz for headroom. A frequency of 150 kHz is far

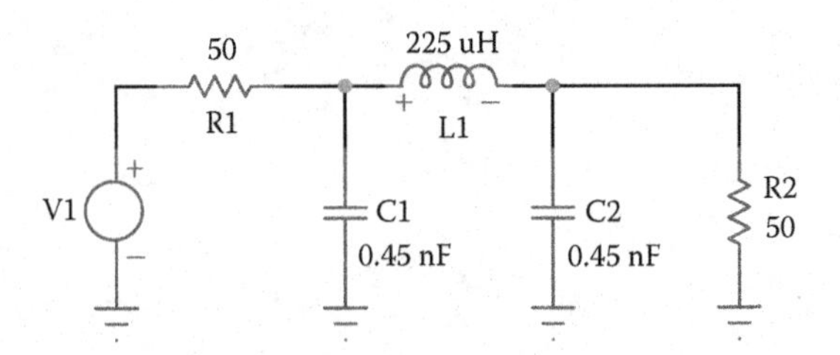

FIGURE 16.1 The π filter designed using equation (16.1).

up in the frequency spectrum and will allow the use of a 50-ohm source and load impedance, so then the design impedance can also be 50 ohms.

What topology should you use? Using the 50 ohms, a π filter could be used due to the input impedance of 50 ohms. This 50-ohm input impedance gives the input capacitor an impedance to work into. How much loss does a π filter give per octave? There are three reactive components, which implies 6 dB per octave per component; therefore, the π filter will provide 18 dB of loss per octave, or 60 dB per decade.

Let's assume that for a π filter, the required loss is 12 dB with a +6-dB margin, or 18 dB. A method follows that provides a *K* value that will solve for the cutoff frequency. Knowing the cutoff frequency (F_0) solves for the component values using equation (16.1). See Figure 16.1. In this application, 18 dB at 150 kHz, the *K* value would be 4 (see Table A.1 in appendix A for 18 dB using a single π), the –3-dB pole-Q, or cutoff frequency F_0, would be 37.5 kHz, and the inductor value equates to 212 μH.

$$L = \frac{R_d}{2\pi F_0} \qquad C = \frac{1}{2\pi F_0 R_d} \tag{16.1}$$

Increase the inductor value to 225 μH and divide this by the impedance squared; $50^2 = 2500$, and the capacitor value is 90 nF. Now check the filter loss with PSpice, or whatever simulation tool is available. The source and load are 50 ohms, the inductor is 225 μH, and the 90 nF is split, with 45 nF each on both the line and load side. If any of the points are still out, add the maximum outage back to the original loss, which was 18 dB. Say the worst point needed another 6 dB. Add this to 18 dB and get 24 dB.

Note that there is no resonant rise in Figure 16.2. If the circuit Q is low enough, there usually is no rise for the single filters—π, L, and T.

The loss requirement was specified as 18 dB, and having adjusted the *L* and *C* values to make them standard, Figure 16.2 shows approximately 19.0-dB loss at 150 kHz.

16.2 Review of the *A* Matrix

Besides the *A* matrix, *Z*, *Y*, *H*, *G*, *B*, and now the scatter parameter (*S*) are used for circuit and system analysis. The advantage of the *A*, or chain, matrix is that each element of a circuit can be chained, or placed in tandem, as the elements of the circuit appear, as in the circuit in Figure 16.3. R_1 and R_3 are series elements, and R_2 and R_4 are shunt elements. Each of these resistors can make up one complete matrix, and each matrix can

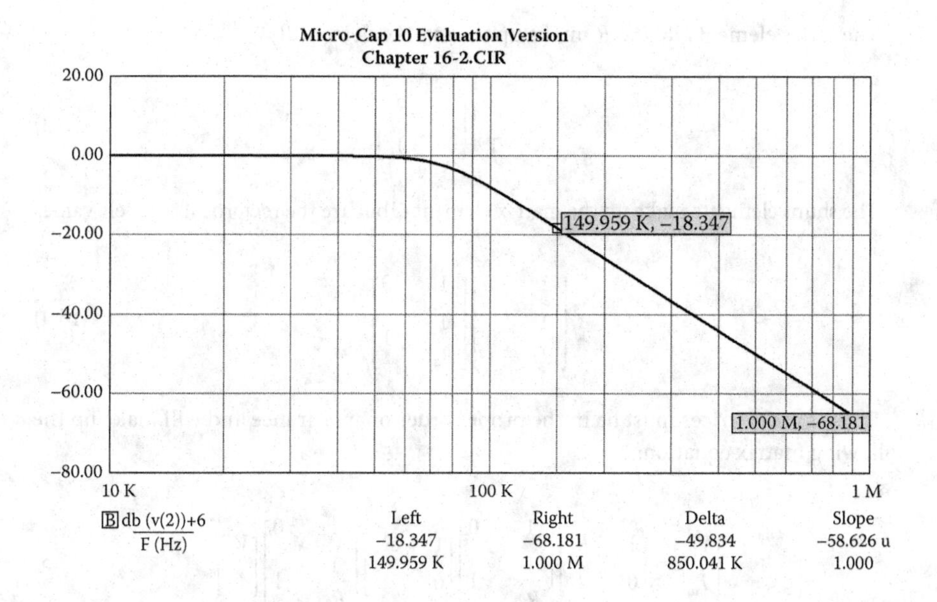

FIGURE 16.2 Loss curve for the EMI filter of Figure 16.1.

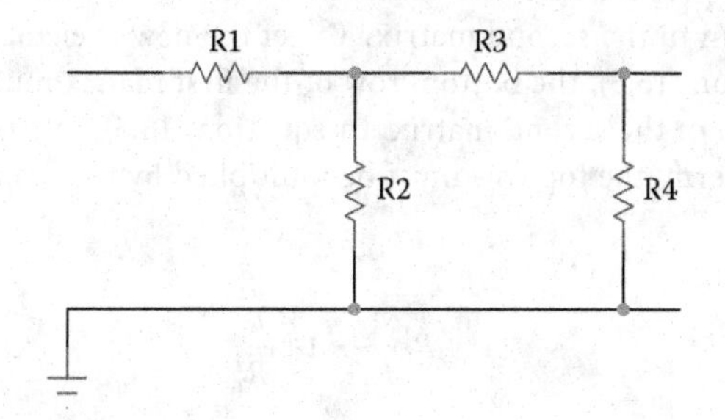

FIGURE 16.3 Simple series and shunt circuit.

be multiplied, or chained, by the following element. Continue through each following element as they appear in the circuit, such as here. This shows four components and four matrices.

$$[R_1][R_2][R_3][R_4]$$

Each individual matrix is formed by four elements, such as

$$\begin{bmatrix} A & B \\ C & D \end{bmatrix} \tag{16.2}$$

The series elements R_1 and R_3 make up the matrix element B

$$\begin{bmatrix} 1 & R_1 \\ 0 & 1 \end{bmatrix} \begin{bmatrix} 1 & R_3 \\ 0 & 1 \end{bmatrix} \tag{16.3}$$

The shunt elements make up the matrix element C but are the reciprocals of their values

$$\begin{bmatrix} 1 & 0 \\ \frac{1}{R_2} & 1 \end{bmatrix} \begin{bmatrix} 1 & 0 \\ \frac{1}{R_4} & 1 \end{bmatrix} \tag{16.4}$$

These four matrices must be in the proper order of appearance and will make up the following matrix equation:

$$\begin{bmatrix} V_{\text{in}} \\ I_{\text{in}} \end{bmatrix} = \begin{bmatrix} 1 & R_1 \\ 0 & 1 \end{bmatrix} \begin{bmatrix} 1 & 0 \\ \frac{1}{R_2} & 1 \end{bmatrix} \begin{bmatrix} 1 & R_3 \\ 0 & 1 \end{bmatrix} \begin{bmatrix} 1 & 0 \\ \frac{1}{R_4} & 1 \end{bmatrix} \begin{bmatrix} V_o \\ I_o \end{bmatrix} \tag{16.5}$$

Note that they are in the proper order of their appearance. To multiply these, the matrix must be in proper order and in the first matrix, the row elements are multiplied by the column in the second matrix. To get the new D element for the first two matrices in equation (16.5), the bottom row of the first matrix must be multiplied by the second column of the second matrix. In equation (16.5), this is $0 \times 0 + 1 \times 1 = 1$. To get the new A term, the top row must be multiplied by first column of the second matrix. This is

$$1^2 + R_1 \frac{1}{R_2} = 1 + \frac{R_1}{R_2}$$

The matrix multiplication of the first two matrices and the last two together yields

$$\begin{bmatrix} V_{\text{in}} \\ I_{\text{in}} \end{bmatrix} = \begin{bmatrix} 1 + \frac{R_1}{R_2} & R_1 \\ \frac{1}{R_2} & 1 \end{bmatrix} \begin{bmatrix} 1 + \frac{R_3}{R_4} & R_3 \\ \frac{1}{R_4} & 1 \end{bmatrix} \begin{bmatrix} V_o \\ I_o \end{bmatrix} \tag{16.6}$$

If all the Rs are equal, the two 2×2 matrices reduce to

$$\begin{bmatrix} V_{\text{in}} \\ I_{\text{in}} \end{bmatrix} = \begin{bmatrix} 2 & R \\ \frac{1}{R} & 1 \end{bmatrix} \begin{bmatrix} 2 & R \\ \frac{1}{R} & 1 \end{bmatrix} \begin{bmatrix} V_o \\ I_o \end{bmatrix} \tag{16.7}$$

and the final matrix multiplication reduces to

$$\begin{bmatrix} V_{in} \\ I_{in} \end{bmatrix} = \begin{bmatrix} 5 & 3R \\ \frac{3}{R} & 2 \end{bmatrix} \begin{bmatrix} V_o \\ I_o \end{bmatrix} \tag{16.8}$$

If the load resistor is also R, then I_o is equal to V_o/R, and the voltage and current follow.

$$\begin{aligned} V_{in} &= 5V_o + 3\frac{RV_o}{R} = 8V_o \\ I_{in} &= \frac{3V_o}{R} + \frac{2V_o}{R} = \frac{5V_o}{R} \end{aligned} \tag{16.9}$$

The input impedance of the resistive pad is then $8R/5$ or $1.6R$. The only difference between the preceding text and the text to follow is that the EMI filters use reactive components, adding j factors, or imaginary components, to the system.

16.3 Transfer Functions

The advantage of the A matrix is that it transfers the output to the input. If there are five A matrices in tandem, each transfers the quantity from the right to the left. This was the style of matrices used years ago for analog servo systems. A water load was transferred to a pump requiring so much torque; the pump transferred this torque load to the electrical motor, which transferred the required electrical load to a control panel and feedback system. The feedback was from the water system. What this EMI system does is to compare the output level to the input level, which is equated to a dB loss for the EMI filter matrices. This section reviews the chain matrix to take advantage of this tool to calculate the insertion loss. These can be chained to form filter modules such as T, π, L, and other filter elements along with their multiples. The matrix includes the load and source impedances, allowing direct calculation of the insertion loss.

Knowledge of this section should allow easy additions by the reader as needed. All the filters listed can be handled with four elements The user can place all of the elements as needed in tandem, and as long as the chain matrix does not run out of room, the matrices can be chained ad infinitum.

Many companies require CISPR specifications, which are easier to meet, such as for the filter in Figure 16.1. The EMI test laboratory or the company's in-house lab will run the proper tests on the unit or system. The filter loss needed is shown by a frequency plot depicting the outages by the peaks above the requirement level line. These require losses such as 60 dB at 150 kHz, as described in section 16.1, but if the system is out of specification at 180 kHz by only 6 dB, the filter must add this loss at that frequency and fix the other outages along the way. Usually, if the lowest problem frequency is fixed, so are the other outages (are of ten within limits), but if not, add this additional loss to the lower frequency loss and recalculate. This is easy to calculate using the appendix A tables and the technique presented in chapter 17.

This is also why it is suggested that people test the system *first* without the filter to determine the loss needed in the filter. If the system is available, then check it without the filter. Now the amount of loss that the filter must provide is known. The system was out by 6 dB at a particular frequency and now is a −2 dB = 8 db filter loss, and so on at other frequencies.

In other cases, it is just a good educated guess by the equipment or power supply designer using equations given in various chapters of this book. The filter designer's goal is to meet the specified needed loss with some additional headroom. From the final matrix of all the combined elements, the loss at a frequency can be calculated with an estimated load and source impedance. For military equipment, the government normally specifies the insertion loss of a filter at 50-ohm load and source impedance in the 220-A specification. The designer can insert these values for any load and source needed. In our case, the designer would pick a filter arrangement, such as a double L, and use some program to find the cutoff frequency needed to get the loss. Again, these programs are used to detect the cutoff frequency giving the needed loss with about 6 dB of headroom. If the cutoff frequency, F_0 is too low, cutting too deep into the frequency passband (15 times the line frequency is suggested for 400 Hz), another stage must be added (here another L). Whatever method you use to iterate between the needed loss and the cutoff frequency is then used again to find the new higher cutoff frequency, and in making sure that the filter is still not too close to the passband as before. Usually, the passband should be at least 15 times the line frequency for 400 Hz for higher loss requirements.

With using the source and load impedances as 50 ohms, we can treat the L and C values as pure reactive elements. Then the *B* and *C* matrix terms are treated as pure reactive. Again, for the *A* matrix, the determinant must always equal 1 even though there are several *j* or *i* factors involved. In solving the matrices, the final matrix would look like equation (16.10).

$$\begin{vmatrix} V_i \\ I_i \end{vmatrix} = \begin{vmatrix} A & jB \\ jC & D \end{vmatrix} \begin{vmatrix} V_o \\ I_o \end{vmatrix} \tag{16.10}$$

In these matrices, the series elements' (here the inductor's) impedance is entered into the B term, while the shunt, or parallel capacitor's impedance, is located in term C as $1/-jX_C = j/-j^2X_C = j/X_C$. Note that the shunt term is the reciprocal value.

16.4 Review of Matrix Topologies

Some engineers use these programs to reverse engineer a filter by changing the component values until the goal of dB loss at a particular frequency is reached. This and most other similar methods give random values of components that yield the needed loss for the 50-ohm test setup, but can do strange things to the passband in the real world (the passband in EMI filters is really only the source frequency: DC, 50 Hz, 60 Hz, 400 Hz, or whatever). This method can also lead to a resonant rise at a low harmonic of the line frequency, create oscillation heat the filter, or present such a low input impedance that the circuit breaker may trip on turn-on. Also, for 400 Hz, this technique can give a serious voltage rise at 400 Hz. There have been cases in which 120 V has produced 126 V at 400 Hz. Customers do not understand what causes this. It simply is power in = power out + some small filter losses. With 126 V × 10 A = 1260 W, filter loss (core and wire losses) can = 50 W or so. The total is 1310, and 1310/120 = 10.92 A in. This is simply changing power for power.

16.5 π Filter

The π filter is used the most often for EMI filters, and this is especially true in applications where losses start around 150 kHz. So this will be handled first. The π filter requires three matrices. The capacitor impedance is in the first and last term, while the inductor impedance is in the middle. It is obvious that the deltas of these three matrices are each equal to 1.

$$\begin{vmatrix} V_i \\ I_i \end{vmatrix} = \begin{vmatrix} 1 & R_S \\ 0 & 1 \end{vmatrix} \begin{vmatrix} 1 & 0 \\ \dfrac{j}{X_C} & 1 \end{vmatrix} \begin{vmatrix} 1 & jX_L \\ 0 & 1 \end{vmatrix} \begin{vmatrix} 1 & 0 \\ \dfrac{j}{X_C} & 1 \end{vmatrix} \begin{vmatrix} V_O \\ I_O \end{vmatrix} \tag{16.11}$$

Note here that the first matrix on the right-hand side is the source impedance. The values of both X_C and X_L are functions of the design impedance, R_d, and K, the normalized frequency. F_0 is the cutoff frequency. L and C are equal to

$$L = \frac{R_d}{2\pi F_0} \qquad C = \frac{1}{2\pi F_0 R_d} \tag{16.12}$$

$$X_C = \frac{-j}{2\pi FC} = \frac{-2j\pi F_o R_d}{2\pi F} = \frac{-jF_o R_d}{F} = \frac{-jR_d}{K} \tag{16.13}$$

Note that the shunt term is the reciprocal of equation (16.13) and is divided by 2 $=jK/2R$ for π.

We are going to substitute V_o/R_L for I_o, and V_i is divided by V_o. Also, all the R values are equal and are 50 ohms. Solve for V_i and get the V_o to V_i ratio.

$$\begin{vmatrix} V_i \\ I_i \end{vmatrix} = \begin{vmatrix} 1 & R_S \\ 0 & 1 \end{vmatrix} \begin{vmatrix} 1 & 0 \\ \dfrac{jK}{2R_d} & 1 \end{vmatrix} \begin{vmatrix} 1 & jKR_d \\ 0 & 1 \end{vmatrix} \begin{vmatrix} 1 & 0 \\ \dfrac{jK}{2R_d} & 1 \end{vmatrix} \begin{vmatrix} V_O \\ \dfrac{V_O}{R_L} \end{vmatrix} \tag{16.14}$$

$$\frac{V_I}{V_O} = 2 - K^2 + jK\left(2 - \frac{K^2}{4}\right) \tag{16.15}$$

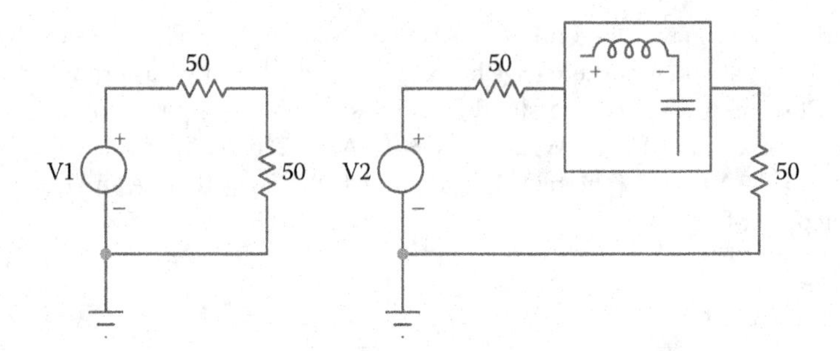

FIGURE 16.4 Insertion loss—with and without the filter.

Square the real and the imaginary sides, add them, and take the square root of the answer.

$$\sqrt{\frac{64+K^6}{16}} = \frac{\sqrt{64+K^6}}{4} \tag{16.16}$$

This voltage ratio is compared with the voltage ratio of the source and load impedances without the filter (Figure 16.4). The loss without the filter is

$$\frac{V_o}{V_i} = \frac{R_L}{R_S + R_L} = 0.5 \tag{16.17}$$

With a typical 50-ohm source and load, the ratio is 2 and the loss is 6 dB.

$$20\log_{10}(0.5) = -6dB \tag{16.18}$$

To get the insertion loss, equation (16.16) is multiplied by equation (16.17). The circuit insertion loss for any single π filter is

$$IL_{dB} = 20Log_{10}\frac{\sqrt{64+K^6}}{8} \tag{16.19}$$

With 12 dB of loss required for the single π filter at 150 kHz, check Table A.1 in appendix A for the corresponding K value for this topology. For 18 dB of loss for a single π filter, search down the right-hand column to find 18.13 dB, and then trace to the value of K in the left-hand column, which is 4.0. This to the sixth power is 4,096.

$$20Log_{10}\frac{\sqrt{64+4096}}{8} = 18.13\,dB \tag{16.20}$$

Equation (16.20) checks out with a value that corresponds to the K value of 4 obtained using the appropriate table in appendix A. The cutoff frequency is 150,000/4 from the table, therefore we have 37.5 kHz.

Example

Design a double π EMI filter for a switched-mode power supply (SMPS) switching at 80 kHz. The test lab states that 30 dB of loss is required at 80 kHz along with other points upstream.

1. In the appropriate dB column of Table A.2 in appendix A for a double π, find 30 dB. The corresponding *K* value is 3.2, and the filter component values will be rounded up to standard part values.
2. Divide 3.2 into 80 kHz, and F_0 is 25,000. This is certainly high enough, even if the source frequency is 400 Hz. The problem here is that there is often a resonant rise well below the cutoff frequency.
3. Calculate L and C.

$$L = \frac{R}{2\pi F_O} = 318 \ \ H.$$

$$\text{From } F_O = \frac{1}{2\pi\sqrt{LC}}$$

$$C = \frac{1}{4\pi^2 F_O^2 L} \triangleq \frac{L}{R^2} = 0.12 \ \ F$$

The *C* value is split—half in front and half in back. The central capacitor is full valued. This design should work, but the current through the *L* may be too much, say 40 A. Try a triple π EMI filter (Table 16.1). See Figures 16.5 and 16.6.

The value at 30.46 dB lists a *K* value of 2.5, and 80 kHz divided by 2.5 gives 32 kHz. As in step 3 above, solve for the inductor and capacitor values. The three inductors are equal to 248 μH and the central capacitors are equal to 0.1 μF, while the first and last are 47 nF.

16.6 L Matrix

A simple L formed with a series inductor followed by a parallel or shunt capacitor would form a small chain matrix such as this. Note that as in the π filter, the small values of

TABLE 16.1 Triple π Excel Spreadsheet

K	Triple π	Sq Root / Divide by 2	20 log dB
2.0	40.00	3.16	10.00
2.1	127.92	5.66	15.05
2.2	357.76	9.46	19.52
2.3	896.96	14.97	23.51
2.4	2065.48	22.72	27.13
2.5	4446.27	33.34	30.46
2.6	9060.09	47.59	33.55
2.7	17635.06	66.40	36.44
2.8	33012.91	90.85	39.17
2.9	59748.64	122.22	41.74

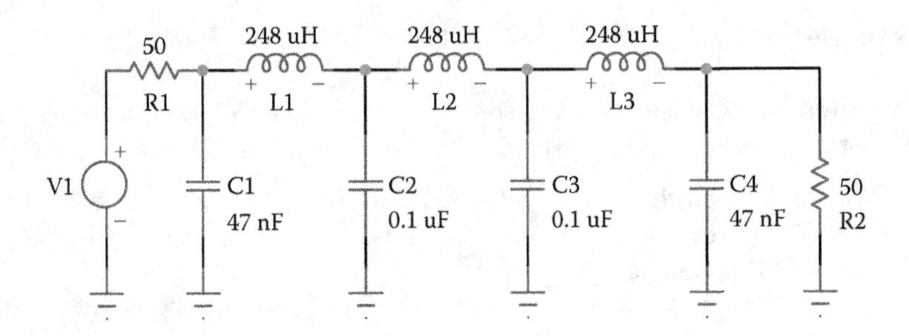

FIGURE 16.5 Triple π filter developed using Table 16.1.

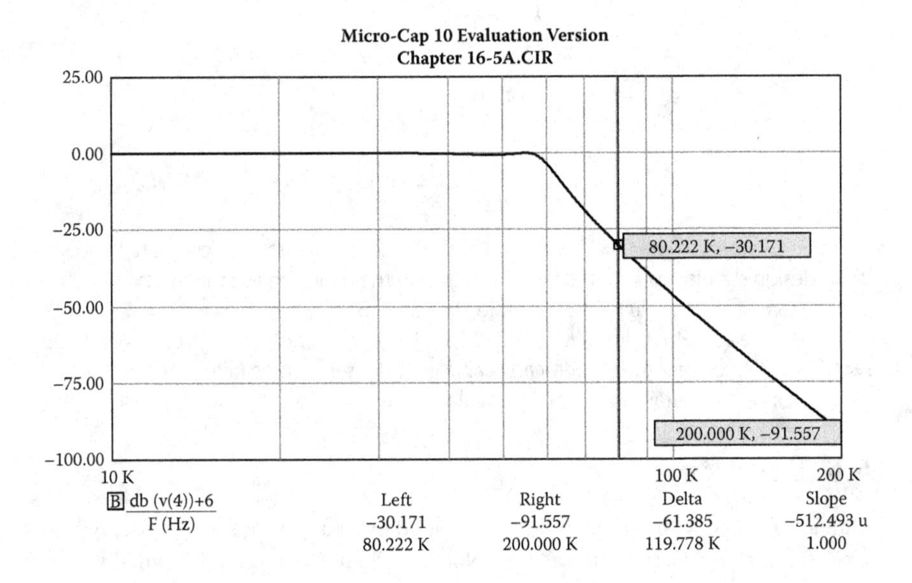

FIGURE 16.6 Triple π filter showing loss at 80 kHz.

equivalent series resistance (ESR) and DC resistance (DCR) are neglected—the components are treated as pure—and the inductor faces the line while the capacitor faces the load. Here, both components are whole. The L filter is often used for EMI filters. Note that the first matrix following the input voltage and current matrix is the source resistance; the second matrix is for the inductor; the third is for the capacitor; and the last is for the output voltage and current.

$$\begin{vmatrix} V_I \\ I_I \end{vmatrix} = \begin{vmatrix} 1 & R_s \\ 0 & 1 \end{vmatrix} \begin{vmatrix} 1 & jX_L \\ 0 & 1 \end{vmatrix} \begin{vmatrix} 1 & 0 \\ \dfrac{j}{X_C} & 1 \end{vmatrix} \begin{vmatrix} V_O \\ I_O \end{vmatrix} \tag{16.21}$$

Following the same procedure of changes as in the π filter

$$\begin{vmatrix} V_I \\ I_I \end{vmatrix} = \begin{vmatrix} 1 & R_s \\ 0 & 1 \end{vmatrix} \begin{vmatrix} 1 & jR_d K \\ 0 & 1 \end{vmatrix} \begin{vmatrix} 1 & 0 \\ \frac{jK}{R_d} & 1 \end{vmatrix} \begin{vmatrix} V_O \\ \frac{V_O}{R_l} \end{vmatrix} \tag{16.22}$$

Solve for V_I/V_0

$$\frac{V_I}{V_O} = (2 - K^2) + j2K \tag{16.23}$$

Square the real and the imaginary terms, add them, and take the square root of this. Then divide the result by 2, the resistor ratios (the loss without the filter). Then take the 20 log to get the dB loss.

What happens if the order is changed? This places the shunt capacitor on the line side and the series inductor facing the load.

$$\begin{vmatrix} 1 & R_S \\ 0 & 1 \end{vmatrix} \begin{vmatrix} 1 & 0 \\ \frac{jK}{R_d} & 1 \end{vmatrix} \begin{vmatrix} 1 & jKR_d \\ 0 & 1 \end{vmatrix} \begin{vmatrix} V_O \\ \frac{V_O}{R_L} \end{vmatrix} \tag{16.24}$$

Multiply the matrices

$$(2 - K^2) + j2K \tag{16.25}$$

$$4 + K^4 \tag{16.26}$$

It is obvious that the solution for the insertion loss is the same for the CL as for the LC (Figure 16.7). For applications with 50-ohm source and load, there is no difference. However, at lower frequencies, the low source resistance shunts the input capacitor of the CL.

Either answer is correct, depending on which element the designer wants facing the line or the load. Most EMI filter designers want the inductor on the line side, especially for MIL-STD-461 specifications or any specification requiring high insertion losses at frequencies near 14 kHz. The reason is that the input inductor adds impedance, giving the following capacitor a higher impedance to work into. A multiple L (double L) can

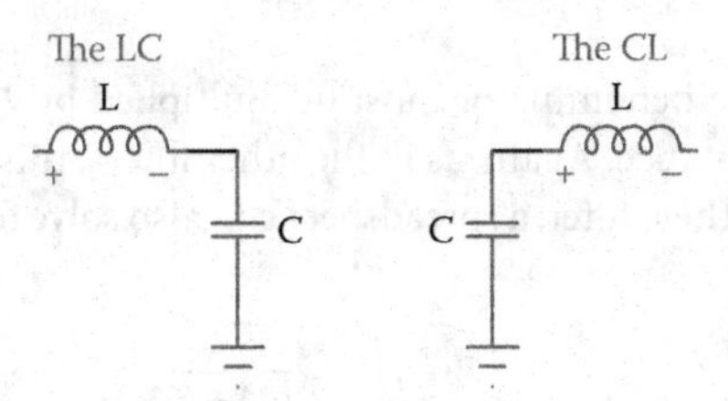

FIGURE 16.7 Comparison of the LC and CL filter matrices.

be generated by the same means. This can be formed by multiplying the two equations together, with the first inductor on the line side.

$$\begin{vmatrix} V_I \\ I \end{vmatrix} = \begin{vmatrix} 1 & R_s \\ 0 & 1 \end{vmatrix} \begin{vmatrix} 1-K^2 & jKR_d \\ \dfrac{jK}{R_d} & 1 \end{vmatrix} \begin{vmatrix} 1-K^2 & jKR_d \\ \dfrac{jK}{R_d} & 1 \end{vmatrix} \begin{vmatrix} V_O \\ \dfrac{V_O}{R_l} \end{vmatrix} \tag{16.27}$$

$$\begin{vmatrix} V_I \\ I \end{vmatrix} = \begin{vmatrix} 1 & R_s \\ 0 & 1 \end{vmatrix} \begin{vmatrix} 1-3K^2+k^4 & jKR_d(2-K^2) \\ \dfrac{jK(2-K^2)}{R_d} & 1-K^2 \end{vmatrix} \begin{vmatrix} V_O \\ \dfrac{V_O}{R_l} \end{vmatrix} \tag{16.28}$$

The determinants in equation (16.28) are still equal to 1, and this can serve to solve for any double L. Solve for V_i.

$$V_I = V_O[2-4K^2+K^4+jK(4-K^2)] \tag{16.29}$$

As normal, divide the source voltage by the output voltage, square the real terms, and square the imaginary terms. Add these two terms, and then the loss ratio is the square root of that answer divided by 2. To get the dB loss, use 20 log (x) of this answer.

16.7 T Filter

This filter is the least used due to the fear of the load-side inductor facing an SMPS. The inductor on the load side may starve the switcher (overly high impedance at the switcher frequency, which creates oscillations). In this filter, the inductor is split, with half of the inductor in series with the line and the other half in series with the load. This is why both inductor matrices are divided by 2. The full matrix of the T is

$$\begin{vmatrix} V_I \\ I_I \end{vmatrix} = \begin{vmatrix} 1 & R_S \\ 0 & 1 \end{vmatrix} \begin{vmatrix} 1 & \dfrac{jKR_d}{2} \\ 0 & 1 \end{vmatrix} \begin{vmatrix} 1 & 0 \\ \dfrac{jK}{R_d} & 1 \end{vmatrix} \begin{vmatrix} 1 & \dfrac{jKR_d}{2} \\ o & 1 \end{vmatrix} \begin{vmatrix} V_O \\ \dfrac{V_O}{R_l} \end{vmatrix} \tag{16.30}$$

$$\frac{V_I}{V_O} = \sqrt{\frac{64+K^6}{16}} \tag{16.31}$$

For insertion loss, the denominator must be multiplied by 2 for the resistive loss of the source and load resistance. Again, as in the other filters, this equation works for any single T filter and, as in the π filter, a spreadsheet can also solve for multiple T filters. The insertion loss is

$$dB = 20\log_{10} 0.5\sqrt{\frac{64+K^4}{16}} \tag{16.32}$$

16.8 Cauer or Elliptic Matrix

The Cauer filter, shown in Figure 16.8, is able to remove a problem frequency, but most often it is used to meet the low-frequency specifications, such as the loss at 20 kHz. In the 461 specifications in which 95 dBμV or so may be needed in the 10-kHz range, a Cauer is often used. F_m is the problem frequency, and R_c is the Cauer resistance (if used) in series with the capacitor. R_c (in Figure 16.8 this is in series with the 0.048-μF capacitor) is there to limit the low impedance of the capacitor leg and lower the Q. For the loss above the problem frequency, F_m, the Cauer looks like a capacitor, and the resistor limits the minimum impedance in this capacitor leg. The impedance of R_d is usually assigned the value of the Cauer resistance, here 50 ohms. However, some leave the Cauer resistor out of the circuit. This is usually employed on the center inductor of a triple-L network. Figure 16.8 shows the central inductor as part of the Cauer. The inductor values are the same as before.

The Cauer capacitor must resonate with the central L at the problem frequency. Because the central inductor is out of the circuit at some frequency above the trouble frequency, any feasible inductor value can be used, but this filter in Figure 16.8 then reduces to a modified double L at the higher frequencies.

$$C = \frac{1}{(2\pi F_m)^2 L} \tag{16.33}$$

Equation (16.33) calculates the value of the Cauer capacitor, and this approaches zero as the frequency increases. Some engineers like the π filter idea better because, above the Cauer resonance, the two capacitors on either side act as the normal value at a center point, which is twice value. In other words, above the Cauer filter, the circuit acts as a normal Quad, or four-section, π filter. Also, the Cauer resistor is not needed because this resistor restricts the parallel combination of two capacitors on either side of the Cauer.

A Quad filter required 100 dB at 20 kHz. The tables in appendix A list the K value at 4.7, which would give 99.35 dB. However, the cutoff frequency calculation of 4,255 Hz is very low and calculates the four inductors at 1870 μH, the three central capacitors at 0.74 μF, and the two end capacitors of 0.38 μF. These inductors are rather large, however but

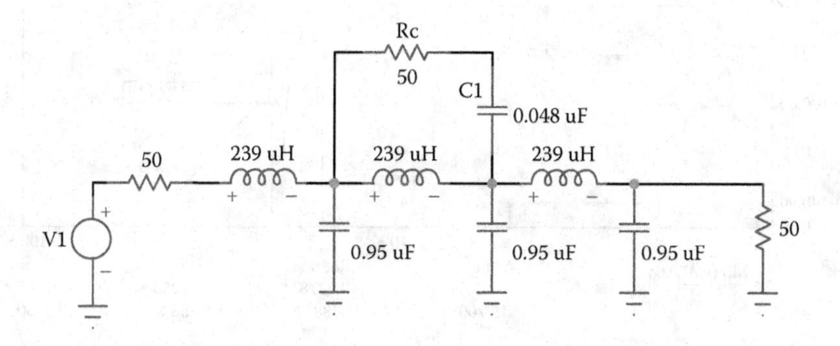

FIGURE 16.8 Center inductor used as part of a Cauer filter.

a Cauer filter can help (Figure 16.9). The Cauer impedance should be 50 ohms, and the resonant frequency has to be close to 20 kHz. The Cauer values were calculated to be 400 μH and 0.16 μF. These are calculated by manipulating

$$50 = \sqrt{\frac{L}{C}}$$

$$C = \frac{1}{2\pi F50}$$

The main concern here is the size of these inductors, even though they have been reduced to 1400 μH. Figure 16.10 shows a resonant rise at 15.7 kHz, a resonant rise at 10.5 kHz, and out of limits at 21.5 kHz by 2.3 dB. Now is the time to bring in the RC shunt.

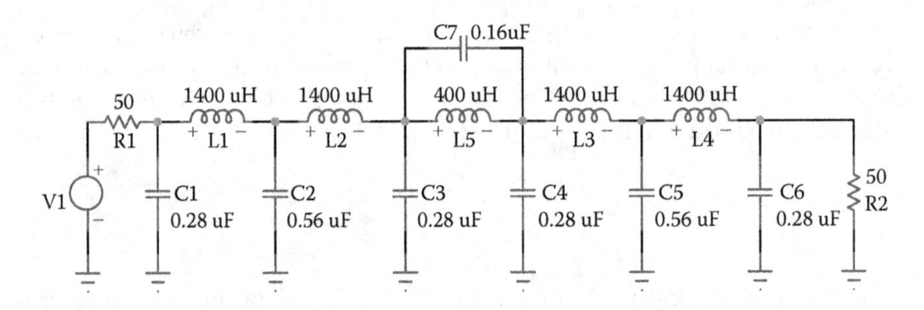

FIGURE 16.9 Quad π filter with a center Cauer.

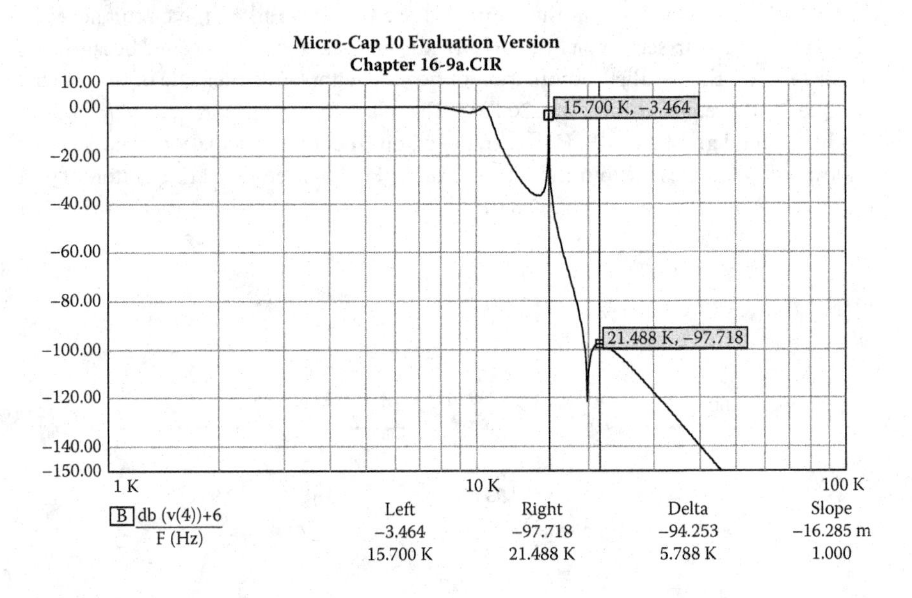

FIGURE 16.10 Quad π filter frequency plot.

It all depends on the current requirement. The higher the current, the more difficult the inductor design is, and the Q is normally low on these large high-current inductors. Figure 16.10 shows a large slope above the Cauer.

16.9 RC Shunt

If the shunt capacitor has added resistance, as with an RC shunt, it is the reciprocal of $R - jX_c$. T his would also be placed in the matrix C term and really makes all the matrix calculations much more difficult. There is an easier way.

$$\begin{bmatrix} A & B \\ [C] & D \end{bmatrix} \Rightarrow \frac{R}{[R^2 + X_C{}^2]} + \frac{jX_C}{[R^2 + X_C{}^2]} \tag{16.34}$$

This RC shunt is used to lower the Q of the filter and correct for a resonant rise and/or a problem related to frequency peak, such as insufficient loss at a particular frequency. To calculate the required capacitor, the reactance must equal the design impedance at the problem frequency. The resistor is typically equal to the design impedance of the filter, and the RC shunt combination is placed across the line. (See components C8 and R3 in Figure 16.11.)

These are normally mounted inboard tied across any of the shunt capacitors where both L and C connect. This filter section can cure several problems at once. Determine the lower resonant rise frequency problem and tie the resultant network across any of the capacitors; the resonant rise will then be attenuated. See Figure 16.12. The loss around 20 kHz above the resonant rise should also be reduced. The circuit Q is reduced to impede any oscillations by the addition of the RC shunt. This is easier to design than the Cauer and is automatically balanced. A 0.35-μF capacitor in series with the design impedance of 50 ohms handles this.

Notice the improvement in the overall curve. The resonant rise is gone, the ring has been greatly reduced and is now down to 35 dB, and the outage at 21.57 kHz now reads 99.99 dB. The self-resonant frequency (SRF) of this RC shunt capacitor need not be the same quality as that of the rest of the filter capacitors unless there are also much higher frequency problems. This is being used primarily to solve low-frequency problems, and

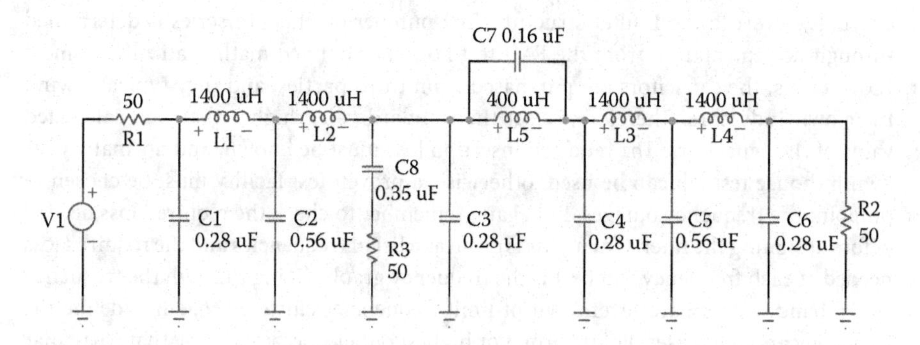

FIGURE 16.11 Addition of RC shunt (R3 and C8) to Figure 16.9.

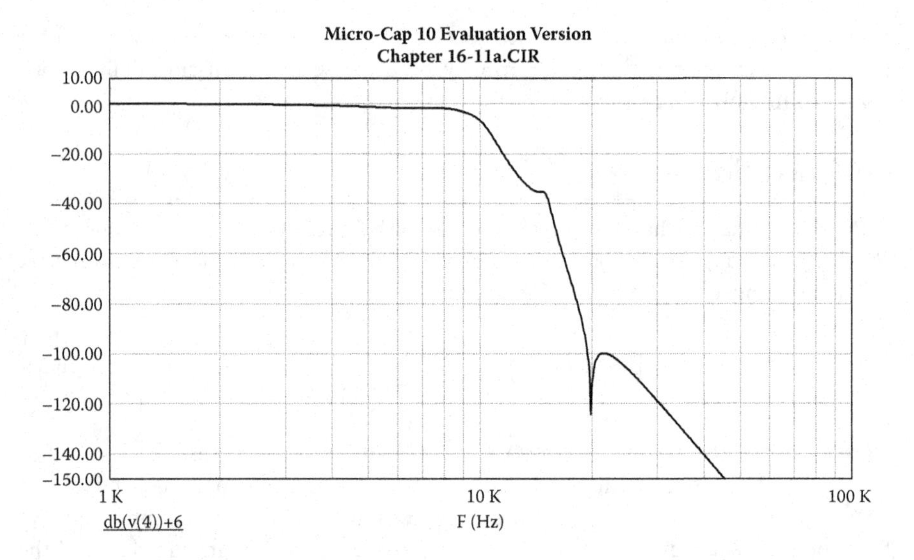

FIGURE 16.12 Frequency plot of Figure 16.11.

the highest frequency in this case is 20 kHz, as just stated. If the SFR is better than the fifth harmonic—100 kHz in the case of this upper frequency—the job will get done. This statement is being made so that the filter designer does not select a capacitor that costs many times more than one that will do the job. The current through this RC capacitor should be minimal. Therefore, a film capacitor is a valid choice.

16.10 Filter Applications and Thoughts

The first step is to find the proper filter type required for the application. For example, a π filter can be used for most applications that will require testing at the 220-A specification (the 50-ohm source, load, and design impedance). They cannot be used where low frequencies are required to filter a large loss unless they are to be tested in a 50-ohm or 220-A system. Make sure that T filters are not used for SMPS due to potentially starving the switcher. This is due to the large inductive reactance at the switcher frequency. If all else fails, revert to the L filter structure. The number of filters in series is determined through documentation from the EMI test house, or defined mathematically. In most EMI houses, the capacitors are purchased from third parties, and they typically wind their own inductors. Therefore, select the standard value higher than the calculated value of the capacitors. The required insertion loss must be known, and normally your own in-house test lab can be used; otherwise, a proven test facility must be chosen to pinpoint the frequency outages. But again, remember to check the required loss needed before designing the filter. Often, this is given as a list of frequencies and the required loss needed at each frequency, but best is the frequency graph. Going through the frequency sweep band, if there are several out-of-limit frequencies clustered together, define the frequency and outage level of the worst or highest outage, because fixing that particular one will ensure that the smaller ones are corrected. The filter must be checked at each of these peak levels to see that they are fixed. If a higher cutoff frequency is required to keep a potential resonant rise away from the incoming line frequency, several options are available.

1. Add a stage to the filter. The component values become smaller and the cutoff frequency increases so the resonant rise will be reduced. The smaller size components have higher SRF and the enclosure can be smaller.
2. Change the topology. Move from an L filter to a π filter, for example.
3. Try an RC shunt tuned to the rise frequency.

The first matrix in the chain is the R_s matrix. This is not part of the filter, as it is from the line or source impedance; however, it does aid in the loss. That is why we deal with insertion loss. The second, third, fourth, and so on, are parts of the filter where the last matrix is a function of the load and also is not part of the filter. These are matrices multiplied together, and the final matrix is solved for the full loss of the filter. As K (the normalized frequency) varies, the loss changes and these can be plotted. This discussion continues in the next section of this chapter.

One condition discussed here is the low-current applications in which operational amplifiers (Op Amps) are used. In certain cases, errors are made when using this application where engineers may substitute active filters to replace passive components in lower current applications. These filters must have *very clean power* or they cross talk, generating the same noise, and in some cases even more noise, than the active filter was meant to cure.

16.11 Single-Phase AC Filter

As an example, an L filter is chosen by the filter designer. The impedances are found to be 50 ohms for source, load, and design; and the line frequency is 400 Hz. The loss needed is 60 dB at 80 kHz. A double L is chosen by the EMI design engineer. Most expect all the losses at the higher frequencies to pass if the lowest passes. This is usually true, but there are exceptions, so all must be checked. This is discussed later; however, the frequency that requires the lowest cutoff frequency determines the cutoff frequency used.

For 60-dB loss in a double L filter, $K = 6.8$ (Table 16.2). Then $F_0 = 11.6$ kHz and the values are

$$L = \frac{50}{2\pi F_O} = \frac{50}{2\pi \times 11600} = 686 \ H \tag{16.35}$$

$$C = \frac{1}{4\pi^2 F_O{}^2 L} \triangleq \frac{L}{R^2} = 0.27 \ F \tag{16.36}$$

See Figures 16.13 and 16.14. Note that there is a potential resonant rise at approximately 17.0 kHz that could require an RC shunt. The series resistor is 50 ohms because the

TABLE 16.2 Double L Excel Spreadsheet

K	Double L Filter	Sq Root / Divide by 2	20 log dB
6.5	2891916.82	850.28	58.59
6.6	3277384.16	905.18	59.13
6.7	3706907.70	962.67	59.67
6.8	4184719.02	1022.83	60.20
6.9	4715381.94	1085.75	60.71
7.0	5303813.00	1151.50	61.23

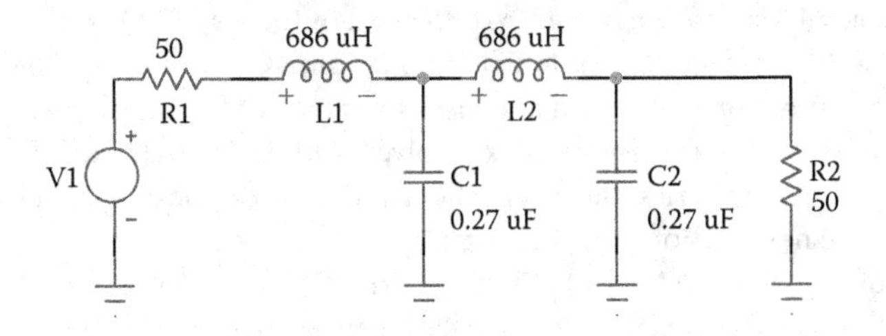

FIGURE 16.13 Double L filter using equations (16.35) and (16.36).

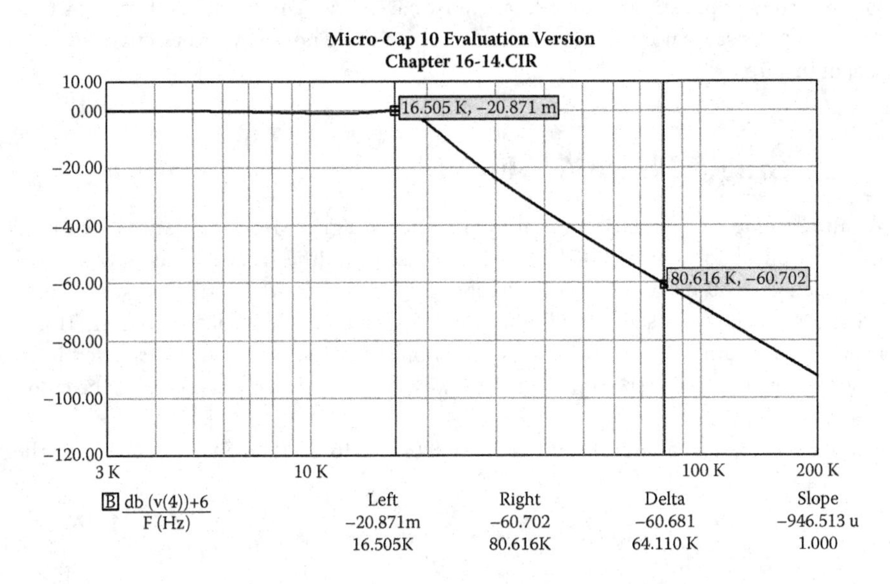

FIGURE 16.14 Frequency plot of Figure 16.13.

source, load, and the design impedances are all 50 ohms. Calculate the value of *C*. The frequency required was reduced to 11.5 kHz for better results.

$$C = \frac{1}{2\pi F_R R_d} = 0.18 \quad F \tag{16.37}$$

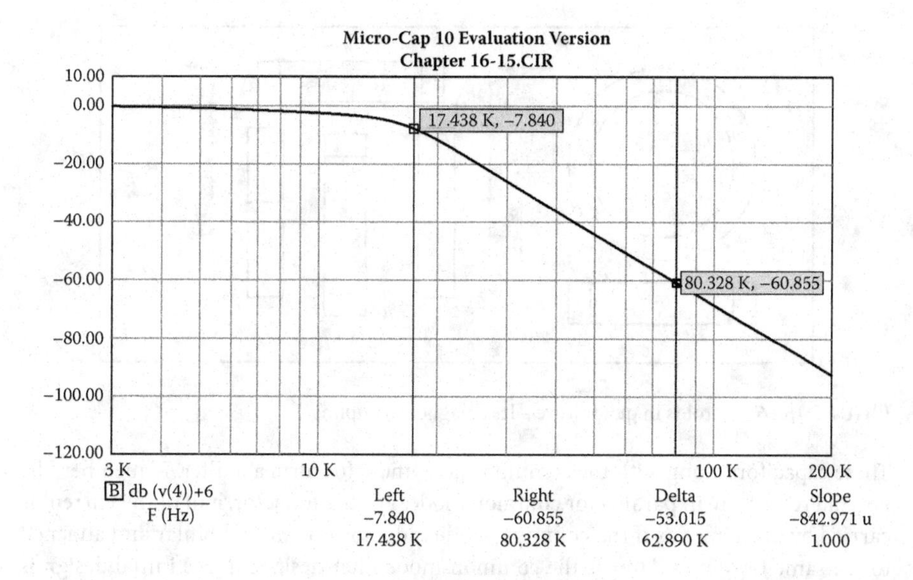

FIGURE 16.15 Curve of Figure 16.14 after adding the RC shunt.

The remaining task is to determine the common mode (Zorro). In many of the figures in this chapter, we show a ground on the return side. In most power service boxes across the world, the return and the ground to the unit under discussion are the same power point. The difference is the value of the two currents. The hot and the return currents should be the same, while the ground should be, in theory, zero. Common mode is seen between any line and ground; however, there is no common mode in these figures. Therefore, to get increased common-mode loss in each of the figures, a balanced filter is needed. If the unit is balanced, a common-mode inductor can be added. But first make sure that there is no "cheat" path around the balance, because if part of the return current is carried on the ground, the system will be unbalanced, and the common-mode inductor will heat and possibly create noise. Advantages of a balanced filter are listed here.

1. An obvious common-mode inductor can be used.
2. The differential inductors can be split—half on both the hot and return. This makes for smaller inductors giving higher SRF, and we can place all the inductors in quadrature (inductors at 90 degrees to each other = less cross talk). See Figure 16.16.
3. The possibility of a smaller enclosure.

The four feed-through capacitors are added, and the value is dependent on the specification. In some specifications, the total capacitance to ground is limited to 0.04 μF. The line-to-line capacitors are out of the circuit along with the RC shunt as far as common mode is concerned. The 686-μH differential-mode inductors are divided by 2, giving 343 μH, and each pair adds 171 μH of common-mode inductance. The total for both sets gives 343 μH for the four which adds this to the Zorro, and provides additional headroom. The common-mode filter is a single π filter. The two input feed-through capacitors are in parallel, and so are the two output load-side feed-through capacitors in parallel.

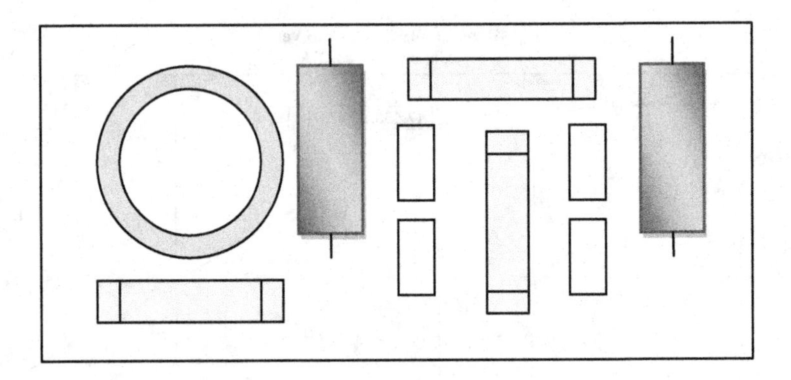

FIGURE 16.16 Toroids in quadrature—less magnetic coupling.

These capacitors along with the common-mode inductor form a π filter. Remember, the hot and return are in parallel for common mode. For the inductor, half of the current is carried by both lines as if the common-mode inductor was wound bifilar and attached to the same terminals. How is the common-mode filter designed? The initial design is carried out almost the same way as the differential-mode filter is designed.

This single-phase filter needs 30 dB at 80 kHz. Looking at the test lab frequency plot, the harmonics of the switcher frequency are visible—both even but mainly odd harmonics of 80 kHz. Figure 16.17 shows that a single π filter is used (so far). Table 16.1 lists for the single π filter for 30 dB a K value of 6.4. Then

$$F_O = \left.\frac{80000}{K}\right|_{K=6.4} = 12{,}500\ Hz$$

$$L = \frac{R}{2\pi F_0} = \frac{50}{2\pi \times 12500} = 636\ \mu H \tag{16.38}$$

$$C = \frac{1}{4\pi^2 F_O{}^2 L} \triangleq \frac{L}{R^2} = 0.27\ \mu F$$

This is not that bad, right? The 12.5 kHz should be good enough for a 400-Hz line frequency, and 640 μH may support the AC line current, but the specification limits the total capacitance value from line to ground to 0.04 μF. Dividing the required capacitance by the limit shows that it is too big by 6.5 times. Common-mode impedance is usually high impedance, especially from the system or equipment and, therefore, does not have to be 50 ohms. So, the common-mode inductor is multiplied by 6.5 = 4200 μH. In Figure 16.17, it is listed at 5 mH and the feed-through capacitors are 0.02 μF. This may be fine depending on the current. If the common-mode inductor is too big for the current, place a central shield across the middle of the enclosure and use two more feed-through capacitors in this shield. Put the RC shunt across the two added feed-through capacitors on either side of the shield. The two common-mode inductors drop in value to 2 mH, one in each half of the enclosure, and each half gets one of the L filters. The capacitors

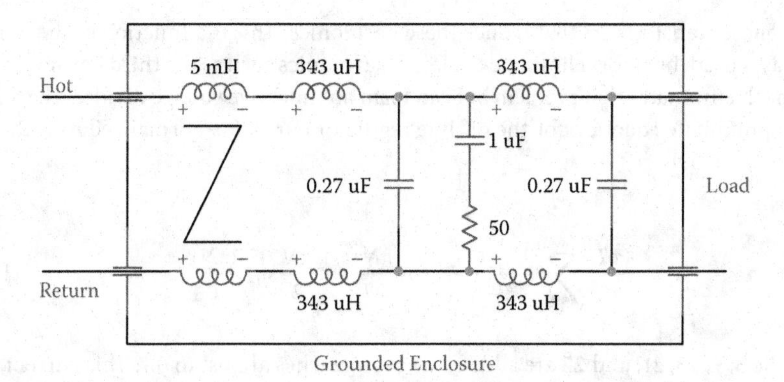

FIGURE 16.17 Modified Figure 16.13 showing RC shunt, Zorro, and feed-throughs.

in the shield are both 0.02 μF. The input and output feed-through capacitors are 0.01 μF, giving a total of 0.04 μF.

All that remains is to design the two different inductors and capacitors, or choose a supplier. This filter may be out slightly when tested but should require only minor changes. Are there high-frequency problems? Add Capcon or ferrite beads. Differential-mode problems at the switcher? Improving the quality of the RC shunt capacitor or increasing its value and/or increasing the inductor values slightly may solve the problem. Do we need more common-mode loss? Increase the Zorro inductance somewhat.

16.12 Three-Phase Filters

There are two types of three-phase filters: the wye (pronounced Y) and the delta. These two types are then divided into two groups. The first is the higher current type, in which each leg is a separate insert filter (each is a different enclosure), with each insert the same type and size. This is very important because of the possibility of installing the smaller neutral filter in any of the main legs. In addition, the third, sixth, ninth, and so on (third-order harmonics), currents add in phase on the neutral wire along with the unbalanced current of the three legs. This harmonic content would be the odd multiples of 3 and some of the even multiples such as 6 or 12. Therefore, the wire size and filter size are the same for the neutral as required for the other three legs.

$$\begin{gathered} A\ \sin(\omega Nt + N0) \\ A\ \sin(\omega Nt + N120) \\ A\ \sin(\omega Nt + N240) \\ 3A\ \sin(\omega Nt) \end{gathered} \tag{16.39}$$

This assumes that the peak harmonic current on all three legs is the same. If not, the term $3A$ is replaced by $(A + B + C)$. For the three-phase neutral, all third-order harmonics add in phase on the ground lead if that harmonic exists. The sixth should be a very low level, if it exists at all. The ninth is strong, especially where the off-line regulator is used as part of the power supplies that are being fed through these filters. Beware of multiphase transformers that reduce these harmonics. This is a function of the number of phases used, but most eliminate the lower harmonics such as the third harmonic. This approach eliminates that problem but can form an inductive voltage divider. The following approximate equation for the off-line regulator current is normalized to 1. Refer to equation (16.39).

$$\frac{4\tau T}{\pi}\sum_{N=1,3,5}^{\infty}\frac{1}{T^2-N^2\tau^2}cos\frac{\pi N\tau}{2T}sin\frac{\pi N}{2}sin\frac{2\pi N\tau}{T} \tag{16.40}$$

If the 3, 9, 15, 21, and 27 are added, the peak reaches almost to 0.7. This current, plus the unbalanced current, should prove that the neutral filter must be the same size as the other three legs. To design the three-phase filter, the voltage and current seen by each filter leg must be known. Find the maximum power required by the load. Divide this by 3 to obtain the power per leg. Divide this answer by the line-to-ground voltage for a wye, or the line-to-line voltage for a delta. In the case of delta, multiply the last answer by the square root of 3.

$$\frac{12.000}{3}=4000 \qquad \frac{4000}{208}=19.231$$

$$19.231\sqrt{3}=33.309 \qquad 33.309\sqrt{2}=47.106$$

If the total load power, taking all the inefficiencies into consideration, is 12 kVA, the power supplied to each line is 4 kVA. If the line-to-line voltage is 208 V, the current is 19.231 times the square root of 3. This equals 33.309 for all the lines A, B, and C. Assuming that no off-line regulators are used, the current peak is 47.106 A. The inductors must not saturate at this current, here 47.1 A.

16.13 Low-Current Wye

The power is from each leg to neutral, and the neutral current is zero if the currents in the three legs are well balanced. But, as mentioned before, the odd third-order harmonics are still present. This is true only for the fundamental of the power line frequency, as discussed previously. In the low-current filter, all of the components are in the same container, and the capacitors are wired from the leg to neutral. This makes for smaller capacitors because of the change in voltage from 208 to 120 V, and this saves money, weight, and volume.

These filters are often called five wire because of the three legs plus the neutral and the ground. The ground must be left intact and unfiltered. We have seen many filters with components in the ground leg from various filter manufacturers, and this is a violation of the electrical code. The ground lead must be intact and not broken. The only exception would be if ferrite beads or toroids were slipped over the wire, leaving the ground lead a solid wire.

The output feed-through capacitors are to case ground, and the ground wire is wired to the case. Also, the Zorro common-mode inductor can be used for common-mode rejection because all parts are within the same enclosure.

The common-mode inductor cancellation of the magnetic field current works as shown in Figure 16.18. The balanced fundamental power frequency is discussed first. Currents in legs A, B, and C (ignoring the common mode) are 120 degrees apart, and these currents generate magnetic fields that cancel in the common-mode inductor. If the currents are different in all three legs, the common mode is no longer balanced and the difference current is carried by the neutral leg. The flow in the neutral must be added to the common-mode inductor, as shown in Figure 16.18. If the system is unbalanced (we know that it is), the difference current flows in the neutral leg of the common-mode inductor and still brings the common-mode inductor back to balance. In these systems, the return winding must be added to the common-mode inductor with the same number of turns that the three other lines have. As far as the harmonics are concerned, whatever harmonic current flows in any leg also flows in the neutral leg and still cancels. So, as far as the power delivered to the load is concerned, the magnetic field of the common mode is neutralized. Any common-mode pulse or signal, either from the load or the line, is attenuated by upsetting this balance. The design process is the same as for a single system. Design the single system and then marry them together. Most of these are tested in the 220-A system with a 50-ohm source, load, and design. Each individual phase is checked one at a time.

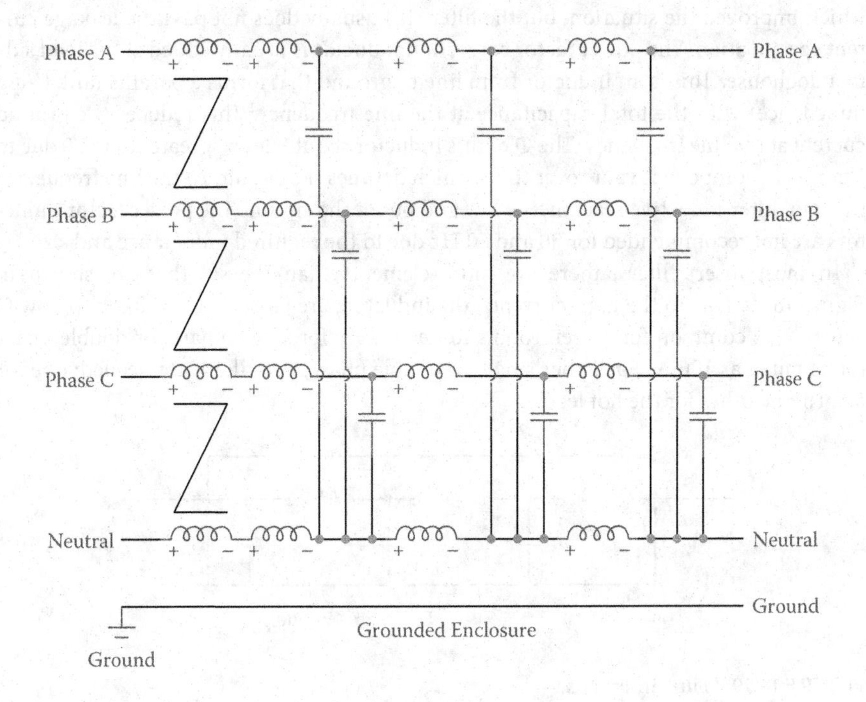

FIGURE 16.18 Three-phase filter with common mode—one enclosure.

16.14 High-Current Wye

These filters become so large and heavy that they must be built using the insert technique. All three legs and the neutral are the same electrically and physically, and all insert filters are installed in a larger enclosure. In this way, if any leg blows, only the failed filter of that leg needs to be replaced. This technique also eases the installation by allowing the outer enclosure to be installed and then the inserts placed in this enclosure one at a time. This is better than hoisting the entire enclosure and contents, which has been done by various contractors. They installed all the inserts, hoisted the full weighted unit, and then dropped it. The entire filter had to be replaced at a premium cost because of time constraints. This was expensive, to say the least.

16.15 Single Insert

All the capacitors in Figure 16.19 are to case ground. In this arrangement, there are no components from line to line, such as capacitors or MOVs. The leakage specification, even with power factor correction coils, is rarely met. There really should not be a leakage current specification for this high-current insert type of filter. This should be known well before the specification is written, and often it is not, and in some cases, it is ignored by some specification writers. In some cases, after the filter was shipped, a request was made that a leakage specification should be met that was not initially provided to the manufacturer. This resulted in power factor correction coils being added, which improved the situation, but the filter still usually does not pass the leakage current specification. The power factor correction inductor is usually added at the load side in a doghouse. This is an inductor from line to ground that forms a parallel tank (high impedance) with the total capacitance at the line frequency. This reduces the ground current at the line frequency. The Q of this inductor should not be greater than 10 due to changes in component value over time, which detunes the circuit. At the line frequency, the total filter inductive reactance is low and out of the circuit. The power factor inductors are not recommended for 50 and 60 Hz due to the required inductance and size.

In most insert filters, there are more elements than the six that are shown in Figure 16.19. Due to the high currents, the inductors are large C cores that have low Q values. It is common for screen rooms to need these for single phase, or double phase for as much as 100 A. For either single or double phase, two filters are needed: one for return, the other for the hot lead.

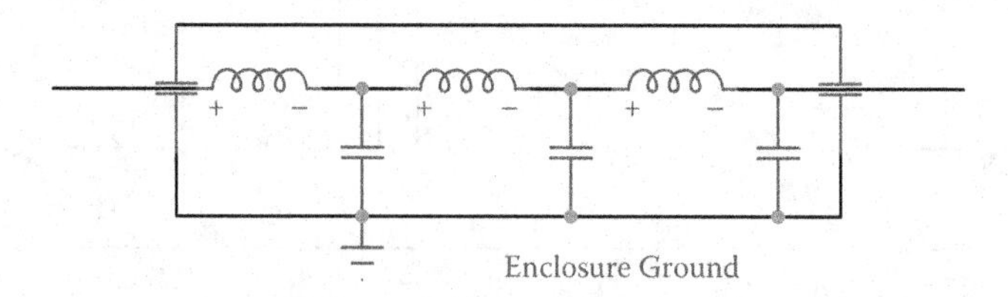

FIGURE 16.19 Filter insert box.

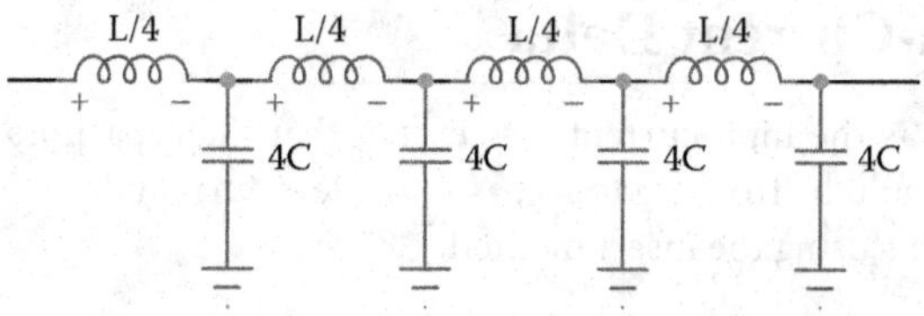

FIGURE 16.20 Equivalent of A, B, C, and return—common mode.

There is, to some degree, a small common-mode influence in the differential inductors, which are in parallel; this also includes the capacitors, which are also in parallel. Each inductor is divided by the number in parallel, and the capacitors multiply in parallel. Also, this assembly lacks the ability to use the Zorro common-mode inductor for common-mode rejection. In the following case, the equivalent circuit is composed of L/4 and 4C, as shown in Figure 16.20. The working impedance is now one-fourth the design impedance, but the cutoff frequency is the same. Also, the amount of common mode is very small.

For common mode, A, B, C, and the return are in parallel; inductors are in parallel divide—or here, four in parallel, and each inductor is divided by 4 and the capacitors add, so again the total common-mode value for capacitance is 4C. These inductors are low valued to keep their size down. So there is some common mode.

$$R_d = \sqrt{\frac{L/4}{4C}} = \sqrt{\frac{L}{16C}} = \frac{1}{4}\sqrt{\frac{L}{C}}$$
$$F_0 = \frac{1}{2\pi\sqrt{(L/4)4C}} = \frac{1}{2\pi\sqrt{LC}} \tag{16.41}$$

16.16 Low-Current Delta

This lower current type is often specified to pass 220 A with the stipulation "to test any one line with the other two legs grounded." This is a good spot for the π type because the line and load impedances are both 50 ohms. The line-to-line capacitors can be shared, giving twice the capacitance to ground. This makes our job easier.

If the A leg is under test in the 220-A system, legs B and C are grounded. Then the capacitors from A to B and A to C are in parallel on both sides of the differential filter inductor. Follow the same procedure for the single phase. If the 220-A specification is all that has to be met, the capacitor values can be divided by 2, making for smaller capacitors and saving money, weight, and volume. If the single π meets all the requirements, the design is finished. If not, add a second π for more loss. Continue with the design as in the single phase. Check that each frequency listed in the specification meets the required loss of the list. If not, the filter may have to be redesigned for more loss or any of the other solutions, depending on the frequency of the problem.

16.17 High-Current Delta

This is the same as the high-current wye except that the capacitors are again tied to ground. These should be from leg to leg, as in the low-current delta. There is no convenient way to do this using the insert method.

16.18 Telephone and Data Filters

These are easy EMI filters to design, and the design program provided here works well. The reason is that the input and output impedance is known, typically 50, 75, 135, or 600 ohms, and the two impedances are always the same. The filter is always balanced; the two inputs are called the tip and ring, and so are the two outputs.

Most telephone filters are 300 ohms from line to ground and 600 ohms line to line—actually tip to ring. These are typically π filters or T filters. The currents are low, and the filter resistance is usually not critical because it is such a small part of the 300 ohms. That is the way to design it. Use the 300 ohms for the source and load, and this will give the filter for the tip and the same for the ring. As an example, a filter requiring 60 dB of loss at 20 kHz matched to 300 ohms would consist of four π filters, as seen in Figure 16.21.

Data filters are not much different; a difference in impedance and the amount of loss required would be about all. However, remember that Figure 16.21 is only half the filter. If this is the filter for the tip input and output, there is another for the ring side, usually in the same can or enclosure.

16.19 Pulse Requirements—How to Pass the Pulse

The quickest way is to obtain the pulse width, take the reciprocal to get the frequency, and multiply this frequency by 10. The EMI filter should then have a cutoff frequency above this frequency. If this impedance is matched, the passband will be very flat. This will pass the fundamental, 3rd, 5th, 7th, and 9th along with some of the 11th harmonic on the slope or sideband.

16.20 The DC-DC Filter

The DC-to-DC type is often a tubular type and is unbalanced. The filter is often a single feed-through capacitor giving just 6 dB per octave loss. In some other filters of this

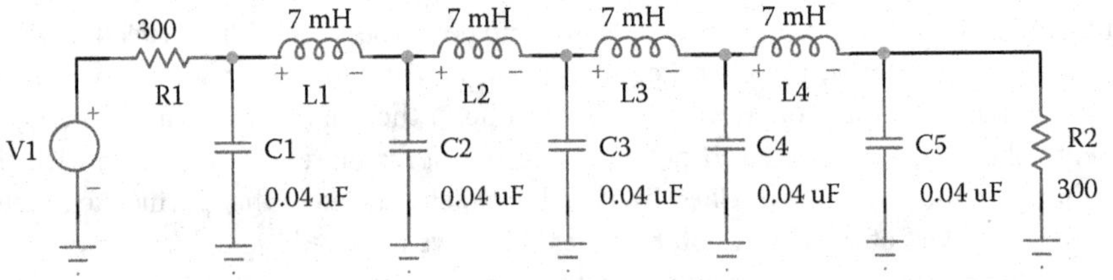

FIGURE 16.21 Telephone filter.

type, inductor(s) are included, making up the L or T type. These last two give 12 or 18 dB per octave.

The output conducting bolt is a snug fit through the capacitor center I_d or hole for the arbor, and this conducting threaded rod is isolated from the outer wall. The capacitor ends are swedged, or soldered, to make two contacts. Tightening the threaded bolt pulls the one swedged capacitor end to the outer wall making the ground connection. The inductor is soldered to the other capacitor end. The outer tube is placed over both the inductor and capacitor, and soldered to the end plate. The inductor is soldered to the other threaded bolt, completing the unit.

16.21 Low-Current Filters

The low-current filter is most often designed to meet the 50-ohm source and load due to these needing to meet higher frequencies at lower loss. Otherwise, they suffer from conditions opposite to those that affect the high-current filter—high inductance and low-valued capacitors.

$$Z = \sqrt{\frac{L}{C}} = \frac{V_{source}}{I_{load}}$$

The inductors get bigger while the capacitors get smaller. One way is to employ RC filters in which the value of *R* should be less than 10% of the minimum load resistance. The disadvantage is that the circuit gives only 6 dB per octave. A better method for low-current filters is to employ active filters. They are small and light and can be designed with many poles, but sometimes the higher frequencies suffer because of the open-loop gain of the Op Amp. It is often overlooked here that the DC feeding Op Amps must be very clean. Therefore, the filter needs a filter, so to speak. There have been cases in which a number of high-impedance lines must be filtered. The ± voltage for each Op Amp must be filtered with an RC filter, or a conventional passive filter, with the capacitor facing the load. The combined voltage feed must be filtered with a passive filter.

$$R = \frac{B^+}{10 \times I_{max}}$$

If the voltage is 12 V and the maximum current is 10 mA, then the maximum value of *R* is 120 ohms. Make *R* 100 ohms, and the capacitive reactance value is also 100 ohms at half the needed cutoff frequency. Note that this gives a 1.0-V drop, and if this is excessive, either the resistance must be lowered or the voltage supply must be increased to allow for it. If 14 kHz is the Op Amp power input cutoff frequency, then at 7 kHz, *C* is equal to 0.22 μF; and round this capacitor up to the next standard value. The same is true for the other supply. If there is a group of these, say 10, then the total B^+ current is 100 mA. This gives an impedance of 120 ohms.

If the Op-amp supply voltage B^+ is 12 V and the maximum current the impedance of the inductor should be 10 times the 120 ohms just calculated, and the impedance of the capacitor should be one tenth this impedance of 120 ohms. A good quality 1-μF capacitor should remove the noise so that one Op Amp does not add noise to the next Op Amp. The inductor removes any of these signals from the main supply.

17

Matrix Applications: A Continuation of Chapter 16

The first step is to find the filter type needed for the application.

If a test specification is in place, this tells the EMI filter designer what specifications the filter must meet, such as MIL-STD-461, CISPR, FCC, or some other. The specification for either the capacitance to ground or the maximum current on the ground lead varies considerably, depending on the function of the equipment. Medical specifications are the hardest to meet, especially if the equipment is attached to the patient. Sometimes the limit is 100 μA total, which means that the individual component specification—say the filter—is limited to 50 μA. If the filter customer has had its product tested for out-of-limit areas using the proper specification, this will tell the product designer the various points along the frequency plot where the unit has failed. Hopefully, this test did not include an unknown or different filter, because the true equipment outage then is really not known. The question is: How much did this filter help in reducing the dB output level, if any?

The 50-ohm specification is based on the typical line and load impedance that both approach, especially in frequencies above 100 kHz. This is not simply a voltage input divided by the current required in the unit. This is also true for the load and is why the LISN (line-impedance stabilization network) is used to match the typical line impedance, and offers a load impedance of 50 ohms to the unit under test. The 50-ohm output of the LISN is from the spectrum analyzer's input impedance, which is 50 ohms and is attached to the LISN. Therefore, the design impedance is 50 ohms, and most off-the-shelf filters are designed and tested at the 50-ohm value. Even the larger filters—high-current screen room type—are often specified to meet the 220-A specification, which is 50 ohms. Therefore, this book mainly addresses the 50-ohm method. What happens if the specification demands a different impedance? Then replace the R value with that impedance value; the equations will still work.

The input voltage and current should be known because the capacitors must meet the proper test voltage. Capacitors must be tested to 4.5 times the peak voltage for AC systems and 2.5 times the peak voltage for DC. The inductors must handle the peak current without saturating. This is one reason why the inductors are designed using approximately half-rated flux density. The higher the flux is in the core, the closer to saturation the inductor becomes. Other things such as arrester issues must be looked into and possibly HEMP. Unfortunately, there is possibly an enclosure size that must be met. This is usually specified by a mechanical engineer and is rarely realistic. This is why electrical engineering should be on board right from the start.

However, the main goal of both chapters 16 and 17 is to define the filter component values. The requirement is for high Q components for better efficiency but low circuit Q. A low circuit Q such as 2 or less tends to stop parasitic oscillations and reduces ringing. So, where do the resistive losses come from? In most cases, resistive losses come from the inductor wire resistance (DCR) and the ESR of the capacitors, but mainly from the line's source resistance (wiring) and any other wiring or resistance in the load side.

17.1 Impedance of the Source and Load

As in chapter 16, these impedances are not part of the EMI filter. This is the output impedance from the line or source, and also the load impedance. But they do add to the dB loss.

$$\begin{bmatrix} V_I \\ I_I \end{bmatrix} = \begin{bmatrix} 1 & R_S \\ 0 & 1 \end{bmatrix} \begin{bmatrix} V_O \\ \dfrac{V_O}{R_L} \end{bmatrix} = \begin{bmatrix} V_O + \dfrac{R_S V_O}{R_L} \\ \dfrac{V_O}{R_L} \end{bmatrix} \tag{17.1}$$

In our case, $R_S = R_L$ and solve for V_I/V_o, then invert to get the voltage ratio. With both source and load being equal, the voltage ratio is simply

$$\frac{R_L}{R_S + R_L} = 0.5 \tag{17.2}$$

then

$$\frac{V_O}{V_I} = 0.5 \rightarrow 20\log_{10}(0.5) = -6\,dB \tag{17.3}$$

Thus, there is 6 dB of loss even without the filter. Therefore, all filter types must add the 6 dB as part of the total loss, or the K equation of the filter must be divided by 2 to get the correct loss. The next section shows the process of generating an equation that gives the loss in dB.

17.2 dB Loss Calculations of a Single π Filter

This filter structure has been selected to show that the two capacitor values are split—half value in front and back. The capacitors are marked in Figure 17.1 as half values. This

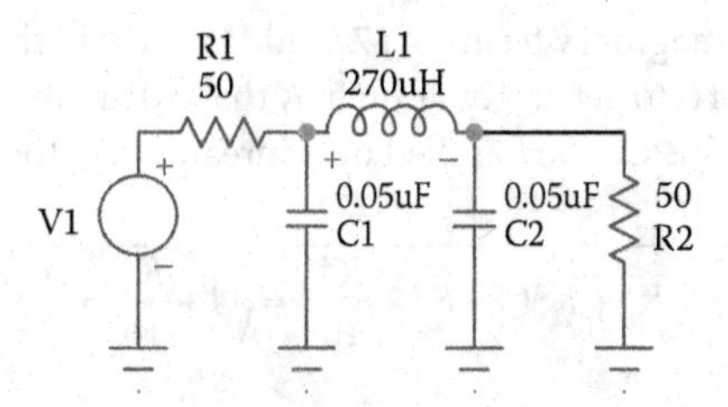

FIGURE 17.1 Single π filter with source and load.

means that their impedances are double. The L is at full value. The jX_L and $-jX_C$ must be calculated.

$$L=\frac{R}{2\pi F_O} \quad jX_L=\frac{j2\pi FR}{2\pi F_O}=\frac{jFR}{F_O}=jKR$$
$$C=\frac{2}{2\pi F_O R} \quad -2jX_C=\frac{-j2\pi F_O R}{2\pi F}=\frac{-2jR}{K} \tag{17.4}$$

The capacitive reactance is placed in term C of the matrix (column 1, row 2), and it is twice the value because the capacitor is half valued. In addition, it is the reciprocal value or inverted. The inductive reactance is placed in B (column 2, row 1)

$$\begin{bmatrix} A & B \\ C & D \end{bmatrix} \tag{17.5}$$

$$\frac{1}{-2jX_C}=\frac{K}{-2jR}=\frac{jK}{2R} \tag{17.6}$$

This requires six matrix equations, as shown in equation (17.7). Starting from the left, the first is the input voltage and current. On the other side of the equal sign, the first matrix is the source matrix. Then the next three matrix terms are the capacitor followed by the inductor and then the last capacitor. These three components make up the π filter. The last matrix is the load. Note that the order of these matrices follows the exact position in which each component is placed, as shown in equation (17.1).

$$\begin{bmatrix} V_I \\ I_I \end{bmatrix}=\begin{bmatrix} 1 & R \\ 0 & 1 \end{bmatrix}\begin{bmatrix} 1 & 0 \\ \frac{jK}{2R} & 1 \end{bmatrix}\begin{bmatrix} 1 & jKR \\ 0 & 1 \end{bmatrix}\begin{bmatrix} 1 & 0 \\ \frac{jK}{2R} & 1 \end{bmatrix}\begin{bmatrix} V_O \\ \frac{V_O}{R} \end{bmatrix} \tag{17.7}$$

The multiplication of the five matrices develops two equations. In the first matrix, V_I develops a voltage ratio of both the input and output that is required to obtain the dB loss of the filter. The second equation calculates the current. The voltage ratio is

$$\frac{V_I}{V_O}=(2-K^2)+jK\left(2-\frac{K^2}{4}\right) \tag{17.8}$$

Square the real and imaginary terms of 17.8, add them, and then take the square root. The reason for all of this is to get an equation in K that calculates the dB loss. As K varies, the dB loss varies, and this can be handled by a spreadsheet. The ratio is

$$\sqrt{4-4K^2+K^4+K^2(4-K^2+\frac{K^4}{16})}=\sqrt{4+\frac{K^6}{16}}=\frac{1}{4}\sqrt{64+K^6} \tag{17.9}$$

But this must be divided by 2 to adjust for the "without filter" loss. The full dB loss is

$$dB=20\log_{10}\left[\frac{1}{8}\sqrt{64+K^6}\right] \tag{17.10}$$

This equation will solve for the dB loss for any single π filter regardless of the problem frequency and required dB if the K value is known. Appendix A lists tables for L, T, and π filters for single, double, triple, and quads. It may be best to use a higher topology if the K value is larger than 10 or more. For example, it may be best to change from a single π filter to a double π filter.

17.3 Example of the Calculations for a Single π Filter

Let's assume that an EMC test revealed the following conclusions. There are outages at 150 kHz, but the highest is at 160 kHz, requiring 12 dB of loss. However, there is a bad section due to the switcher frequency of 250 kHz and the harmonics of the 250 kHz. In addition, 250 kHz requires 25 dB of loss. The question is, "What is the cutoff frequency?" Add at least 3 dB to both losses, so 160 kHz would require 15 dB and 250 kHz would require 28 dB. The K value for 160 kHz is 3.6, and for 250 kHz it is 5.9.

The cutoff frequency using 160 kHz is divided by 3.6 and gives 44.4 kHz, and 250 kHz is divided by 5.9 and gives 42.4 kHz. The lower frequency must then be the cutoff frequency, and this comes from the 250-kHz outage. Now the individual component values can be calculated using 42.4 kHz and the 50-ohm source and load.

$$L=\frac{R}{2\pi F_o}=187.7\,\mu H \tag{17.11}$$

$$\text{From } F_O=\frac{1}{2\pi\sqrt{LC}}$$

$$C=\frac{L}{R^2}\triangleq\frac{187.7}{2500}=75nF \tag{17.12}$$

The capacitor is split, with half in front on the line side and half on the load side. Also, the designer would not use these values. The inductor would be on the order of 200 μH and the capacitors at approximately 68 nF. But for now, we can use the values calculated in equations (17.11) and (17.12).

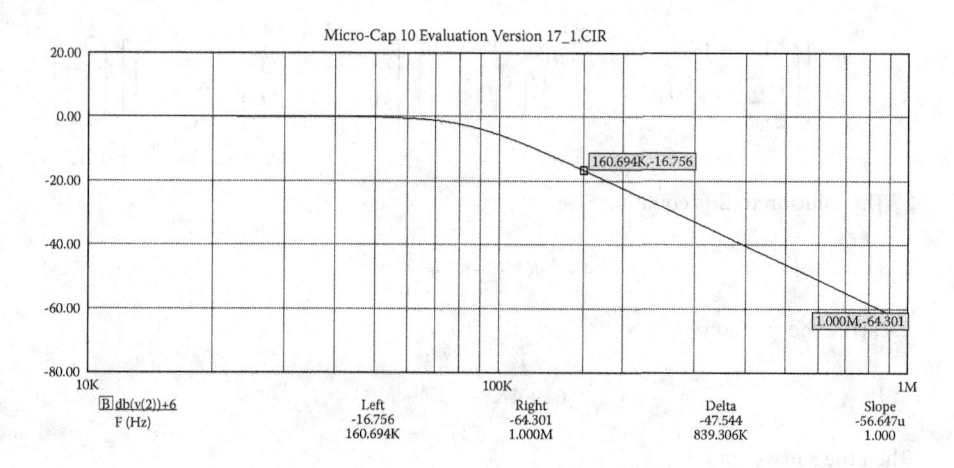

FIGURE 17.2 Plot of the single π filter shown in Figure 17.3.

FIGURE 17.3 Single π filter used in example.

From inspection of Figure 17.2, note that the loss at 160 kHz is approximately -16 dB, and at 250 kHz the loss is over 28 dB, where all the 250-kHz harmonics are sufficiently attenuated as a function of the frequency magnitude slope. This frequency plot has not included the -6 dB loss that would be attributed to the 50 ohms source and load in order to illustrate the actual loss of the filter alone. Also note that Figure 17.3 is using calculated values for the capacitor which are not practicable. The circuit should be simulated with a nearest preferred value for the design capacitance assuming a split value of the calculated 75 nF. We could use two 39 nF capacitors for example. The inductance can be either recalculated, or wound to give as near 187 μH as possible. Either way, it is important to maintain the same pole-Q -3dB frequency.

17.4 Double π Filter: Equations and dB Loss

Some of the equations in this section were defined using Mathcad where the imaginary terms are denoted with the letter 'i'. This is equivalent to 'j' in all cases as used throughout the book. The equation below represents the full matrix for a double π filter. Note the center capacitor as compared to the two end capacitors. Two half capacitors in the center are "married," giving a whole-capacitor equivalent.

$$\begin{pmatrix}1 & R\\0 & 1\end{pmatrix}\cdot\left[\begin{pmatrix}1 & 0\\\frac{i\cdot K}{2\cdot R} & 1\end{pmatrix}\cdot\begin{pmatrix}1 & i\cdot K\cdot R\\0 & 1\end{pmatrix}\cdot\begin{pmatrix}1 & 0\\\frac{i\cdot K}{R} & 1\end{pmatrix}\right]\cdot\left[\begin{pmatrix}1 & i\cdot K\cdot R\\0 & 1\end{pmatrix}\cdot\begin{pmatrix}1 & 0\\\frac{i\cdot K}{2\cdot R} & 1\end{pmatrix}\right]\cdot\begin{pmatrix}1\\\frac{1}{R}\end{pmatrix} \tag{17.13}$$

The solution to this equation is

$$4+\frac{1}{4}K^6-\frac{1}{4}K^8+\frac{1}{16}K^{10} \tag{17.14}$$

Clearing the fraction

$$\frac{1}{16}(64+4K^6-4K^8+K^{10}) \tag{17.15}$$

Then the square root

$$\frac{1}{4}\sqrt{(64+4K^6-4K^8+K^{10}} \tag{17.16}$$

Divide this by 2 for the source and load, and calculate the dB loss

$$dB=20\log_{10}\left[\frac{1}{8}\sqrt{64+4K^6-4K^8+K^{10}}\right] \tag{17.17}$$

This will give the dB loss for any double π filter if the K value is known.

17.5 Triple π Filter: Equations and dB Loss

Note the two central capacitors in the following equation, again whole valued.

$$\begin{pmatrix}1 & R\\0 & 1\end{pmatrix}\begin{bmatrix}\frac{9}{2}\cdot i^2\cdot K^2+3\cdot i^4\cdot K^4+\frac{1}{2}\cdot i^6\cdot K^6+1 & 4\cdot i^3\cdot K^3\cdot R+i^5\cdot K^5\cdot R+3\cdot i\cdot K\cdot R\\\frac{1}{4}\cdot i\cdot K\cdot\frac{3+i^2\cdot K^2}{R}\cdot(4+5\cdot i^2\cdot K^2+i^4\cdot K^4) & \frac{1}{2}\cdot(2+i^2\cdot K^2)\cdot(4\cdot i^2\cdot K^2+i^4\cdot K^4+1)\end{bmatrix}\cdot\begin{pmatrix}1\\\frac{1}{R}\end{pmatrix} \tag{17.18}$$

Solution of the equation above yields

$$\begin{pmatrix}9\cdot i^2\cdot K^2+6\cdot i^4\cdot K^4+i^6\cdot K^6+2+6\cdot i\cdot K+\frac{35}{4}\cdot i^3\cdot K^3+3\cdot i^5\cdot K^5+\frac{1}{4}\cdot i^7\cdot K^7\\\frac{1}{4}\cdot\frac{12\cdot i\cdot K+19\cdot i^3\cdot K^3+8\cdot i^5\cdot K^5+i^7\cdot K^7+18\cdot i^2\cdot K^2+12\cdot i^4\cdot K^4+4+2\cdot i^6\cdot K^6}{R}\end{pmatrix} \tag{17.19}$$

Then get the voltage ratio, take the square root of it, and divide by 2 to compensate for the load and source.

$$dB=20\log_{10}\left[\frac{1}{8}\sqrt{64+9K^6-24K^8+22K^{10}-8K^{12}+K^{14}}\right] \tag{17.20}$$

TABLE 17.1 Single π Filter

K	$64 + K^6$	$\sqrt{x}$	$x/2$	$20 \log_{10}$ (dB)
3.0	49.56	7.04	3.52	10.93
3.1	59.47	7.71	3.86	11.72
3.2	71.11	8.43	4.22	12.50
3.3	84.72	9.20	4.60	13.26
3.4	100.55	10.03	5.01	14.00
3.5	118.89	10.90	5.45	14.73
3.6	140.05	11.83	5.92	15.44
3.7	164.36	12.82	6.41	16.14
3.8	192.18	13.86	6.93	16.82
5.5	1734.04	41.64	20.82	26.37
5.6	1931.56	43.95	21.97	26.84
5.7	2147.53	46.34	23.17	27.30
5.8	2383.29	48.82	24.41	27.75
5.9	2640.28	51.38	25.69	28.20
6.0	2920.00	54.04	27.02	28.63
6.1	3224.02	56.78	28.39	29.06
6.2	3554.01	59.62	29.81	29.49

This equation will solve for the dB loss for any triple π filter if the K value is known.

The K values can be looked up in the tables in appendix A. Pick a topology, such as a double π filter; then follow down the dB column to the loss required, and read the K value on that row. Divide the K value into the outage frequency or a frequency where you know what loss is required to get the cutoff frequency, then solve for L and C. Regardless of the topology, say the K value is 5 and the frequency requiring the loss is 170 kHz. The calculation would be as follows:

$$F_O = \frac{170000}{5} = 35\,kHz$$

$$L = \frac{50}{2\pi \times 35000} = 227\,\mu H \tag{17.21}$$

$$C = \frac{1}{4\pi^2 F_O^2 L} \triangleq \frac{L}{R^2} = 90\,nF$$

Due to the fact that we are dealing with a triple π filter, all three inductors are 227 μH and probably rounded up to 230 μH. However, for the end capacitors—the line side and load side are half valued—some standard value above 45 nF is acceptable. The two central capacitor values would be 90 nF or rounded up to possibly 0.1 μF.

18

Network Analysis of Passive LC Structures

This chapter presents frequency-domain analysis of EMI LC filter structures in terms of network impedances, transfer functions, stability, and insertion loss. This chapter is also a prerequisite for chapter 19, which provides discussion on EMI filter design process and presents formal EMI filter design methods using various analysis techniques.

18.1 Lossless Networks

In almost all cases, EMI filters are made up from one or more LC structures and should include carefully placed damping dQ networks to reduce the circuit Q for each complex pole pair. Lossless networks are those that are constructed of purely reactive elements, so that no losses are incurred in the network itself. In reality, these reactive elements have equivalent series resistance and, therefore, the filter will incur small losses due to either inductor current or capacitor ripple current. In order to provide a greater slope or roll off, it is possible to cascade several low-pass filter sections. When this is done, the filter elements from adjacent sections may be combined.

For example, if two T-section filters are cascaded and each T section has a 1-μH inductor in each leg of the T, these may be combined in the adjoining sections and a 2-μH inductor used. A cascade π filter may also use this approach, where the capacitance between each series inductor is the sum of the two capacitors. In some EMI applications, it is acceptable to just cascade two or more sections, but this will make the design of a filter more complex, particularly with controlling impedances between adjacent sections and in defining an accurate pole-Q frequency. Figure 18.1 shows three simple filter sections—L, T, and π—constructed symmetrically with impedances Z_1 and Z_2. An EMI filter presents a mismatch in impedance between the line and load to which it is connected. The low-pass configuration of this device has series inductors, high impedance (Z_1) with increasing frequency, and shunting capacitors from line-to-line and line-to-ground, which present low impedance (Z_2) with increasing frequency. To summarize, the L network consists of a series impedance Z and a shunt admittance Y. The combination of these components results in a circuit cascade of either T or π structures with high

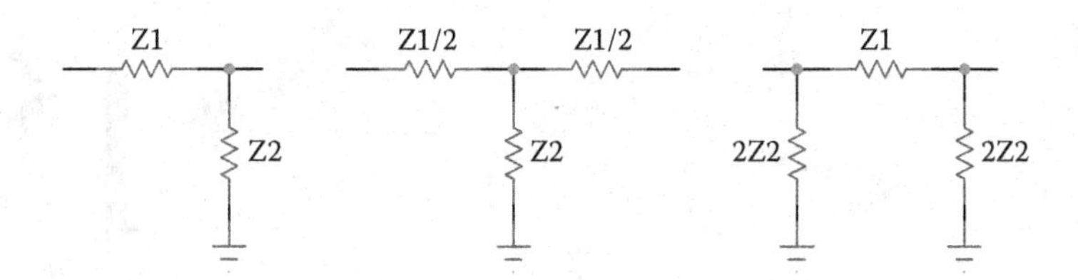

FIGURE 18.1 LC filter sections (a) L, (b) T, (c) π.

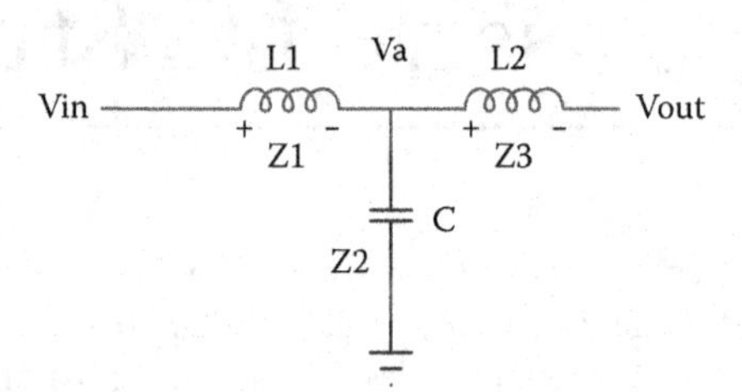

FIGURE 18.2 LC filter T section with series resistance.

series impedance. This circuit structure impedes the flow of the high-frequency content and effectively shorts it to ground through the capacitors. If we consider Figure 18.1a, we have impedances Z_1 and Z_2, which form an L section and may be defined as follows: $Z_1 = j\omega L$ and $Z_2 = 1/j\omega C$.

In a practical EMI filter solution, the circuit will include parasitic elements in the form of ESR, or the equivalent series resistance within the inductors. This resistance is intrinsic to the inductive element and largely dictated by the wire size (AWG) and wire length. An undamped LC filter will have a much higher Q-factor when the DC series resistance is small. Figure 18.2 captures a T filter network. For practical purposes, we shall assume the following impedance functions:

$$Z_1(\omega) = j\omega L_1 \qquad Z_2(\omega) = \frac{1}{j\omega C} \qquad Z_3(\omega) = j\omega L_2 \tag{18.1}$$

18.2 Network Impedances Using *Z* Parameters

The EMI filter may be regarded as a two-port network, and the use of either the *Y* or *Z* parameter is most common in allowing simple analysis of circuit input and output impedances based upon circuit voltages and currents. This approach is not always used in EMI filter design, as we are not really interested in matching impedances; however, it will present a useful summary of how the input and output impedances of these LC networks are defined.

The *Z* parameters, or open-circuit parameters, relate to the output currents from their ports to their input voltages (Figure 18.3). The impedance matrix is given by

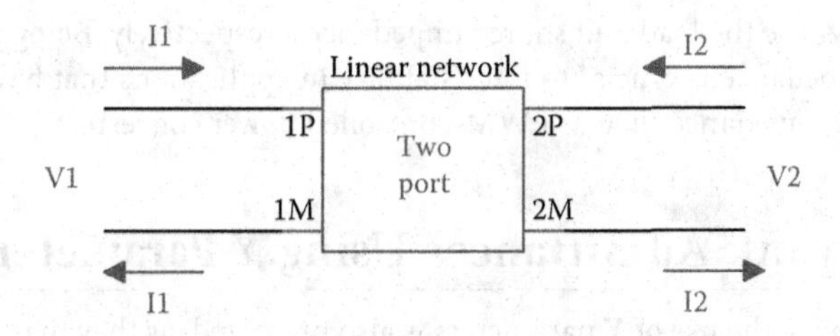

FIGURE 18.3 General two-port network.

$$\begin{pmatrix} V_1 \\ V_2 \end{pmatrix} = \begin{pmatrix} z_{11} & z_{12} \\ z_{21} & z_{22} \end{pmatrix} \begin{pmatrix} I_1 \\ I_2 \end{pmatrix} \Leftrightarrow \begin{pmatrix} V_1 \\ V_2 \end{pmatrix} = \begin{pmatrix} z_{11}I_1 & z_{12}I_2 \\ z_{21}I_1 & z_{22}I_2 \end{pmatrix} \tag{18.2}$$

where

$$\begin{matrix} z_{11} = \left.\dfrac{V_1}{I_1}\right|_{I_2=0} & z_{12} = \left.\dfrac{V_1}{I_2}\right|_{I_1=0} \\ z_{21} = \left.\dfrac{V_2}{I_1}\right|_{I_2=0} & z_{22} = \left.\dfrac{V_2}{I_2}\right|_{I_1=0} \end{matrix} \Rightarrow \begin{matrix} V_1 = z_{11}I_1 + z_{12}I_2 \\ V_2 = z_{21}I_1 + z_{22}I_2 \end{matrix} \tag{18.3}$$

where

z_{11} = open circuit input impedance (port 1 driving point impedance)
z_{12} = open circuit transfer impedance from port 1 to 2
z_{21} = open circuit transfer impedance from port 2 to 1
z_{22} = open circuit output impedance (port 2 driving point impedance)

If we look at Figure 18.2, we can define the impedance matrix as follows:

$$[Z] = \begin{bmatrix} z_1 + z_2 & z_2 \\ z_2 & z_2 + z_3 \end{bmatrix} \tag{18.4}$$

The Z parameter matrix for Figure 18.2 is therefore

$$[Z] = \begin{bmatrix} z_{11} = j\omega L_1 + \dfrac{1}{j\omega C} & z_{12} = \dfrac{1}{j\omega C} \\ z_{21} = \dfrac{1}{j\omega C} & z_{22} = \dfrac{1}{j\omega C} + j\omega L_2 \end{bmatrix} \tag{18.5}$$

There are two impedance relationships within the Z parameter matrix that can be used to analytically define both the input and output impedances based upon the inclusion of the source and load impedances. These are defined in equation (18.6).

$$Z_{in} = z_{11} - \frac{z_{12}z_{21}}{z_{22} + z_L} \qquad Z_{out} = z_{22} - \frac{z_{12}z_{21}}{z_{11} + z_S} \tag{18.6}$$

where z_L and z_S are the load and source impedances, respectively. Being able to predict the output impedance is crucial to filter stability in applications that have incremental negative input impedance such as PWM-controlled power converters.

18.3 Network Admittances Using *Y* Parameters

For completeness, the use of *Y* parameters is also presented, as they may be used where a transfer function is based upon the reciprocal of impedance, or admittance *Y*. The *Y* parameters or short-circuit parameters relate the output voltages from the ports to their input currents. We can illustrate this as follows:

$$[V]=[Z].[I]\rightarrow[I]=[Z]^{-1}.[V]\Rightarrow[Y]=[Z]^{-1} \tag{18.7}$$

$$\begin{matrix} y_{11}=\left.\frac{I_1}{V_1}\right|_{V_2=0} & y_{12}=\left.\frac{I_1}{V_2}\right|_{V_1=0} \\ y_{21}=\left.\frac{I_2}{V_1}\right|_{V_2=0} & y_{22}=\left.\frac{I_2}{V_2}\right|_{V_1=0} \end{matrix} \Rightarrow \begin{matrix} I_1=y_{11}V_1+y_{12}V_2 \\ I_2=y_{21}V_1+y_{22}V_2 \end{matrix} \tag{18.8}$$

Using *KCL*, we can write the following for Figure 18.4:

$$I_1=V_1Y_1+(V_1-V_2)Y_2=V_1(Y_1+Y_2)-V_2Y_2$$

$$I_2=V_2Y_3+(V_2-V_1)Y_2=-V_1Y_2+V_2(Y_2+Y_3)$$

From these, we can write the *Y* parameters

$$[Y]=\begin{bmatrix} Y_{11} & Y_{12} \\ Y_{21} & Y_{22} \end{bmatrix}=\begin{bmatrix} Y_1+Y_2 & -Y_2 \\ -Y_2 & Y_2+Y_3 \end{bmatrix} \tag{18.9}$$

From Figure 18.4, we can write the admittance matrix as follows:

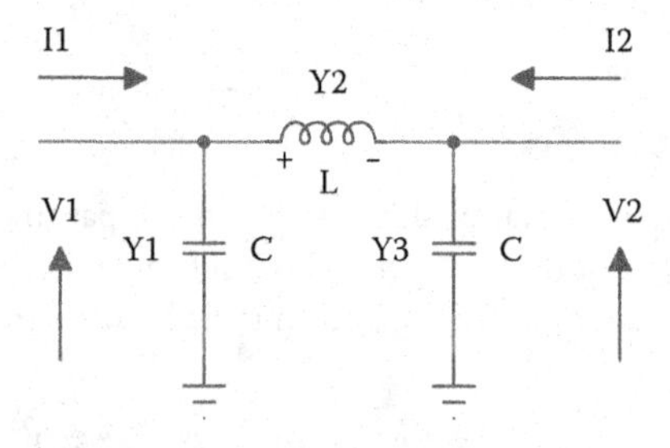

FIGURE 18.4 LC filter π section.

$$[Y] = \begin{bmatrix} j\omega C + \frac{1}{j\omega L} & -\frac{1}{j\omega L} \\ -\frac{1}{j\omega L} & j\omega C + \frac{1}{j\omega L} \end{bmatrix} \quad (18.10)$$

18.4 Transfer Function Analysis—*H*(*j*ω)

The transfer function of Figure 18.5 with both source and load impedances included is often referred to as a doubly terminated LC structure, and may be defined using a process of voltage division

$$\frac{v_o}{v_i}(j\omega) = H(j\omega) = \frac{(z_L + j\omega L_2) \| \frac{1}{j\omega C_1}}{\left[(z_L + j\omega L_2) \| \frac{1}{j\omega C_1}\right] + (j\omega L_1 + z_S)} \cdot \frac{z_L}{z_L + j\omega L_2} \quad (18.11)$$

Setting ω = 0, the DC gain is obtained, since the inductors appear short circuit and the capacitor appears open circuit. For the unique case where the source (z_S) and load (z_L) are resistive and both equal to the filter characteristic impedance, the circuit Q is minimum and the DC gain is

$$\left.\frac{v_o}{v_i}\right|_{\omega=0} = \frac{R_L}{R_L + R_S} \Rightarrow 20\log_{10} 0.5 = (-6dB)$$

$$L_1 = L_2 = 0.5L, C = C, R_L = R_S = \sqrt{\frac{L}{C}}$$

Most practical EMI filters are based upon simple *L* networks (*LC*) and are connected between the source and load. Most often, the load is the input to a power supply, motor controller, or some other equipment that is power supply controlled. In these particular cases, the filter output impedance has to be very low. If we consider Figure 18.6, the

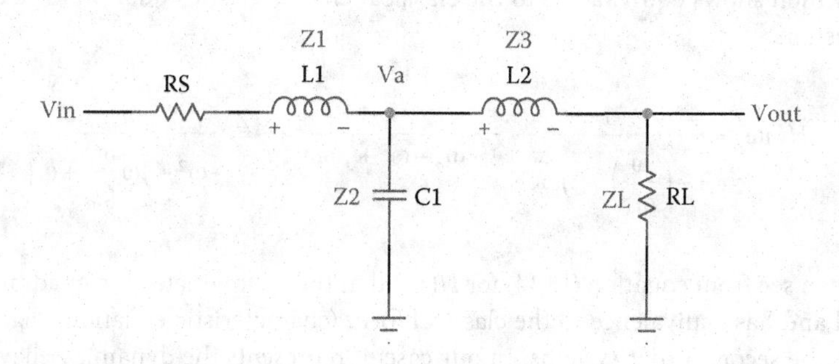

FIGURE 18.5 LC filter T section.

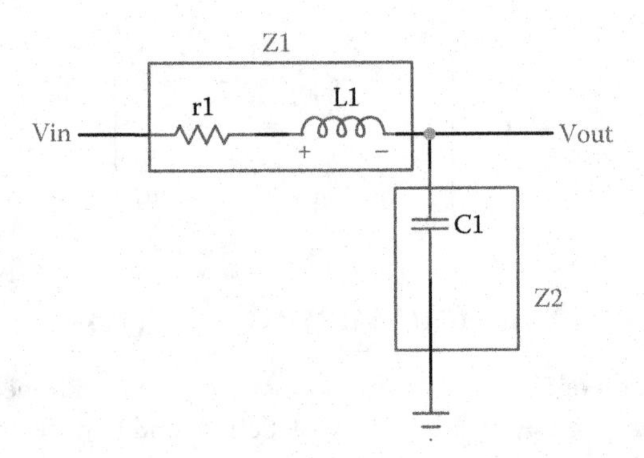

FIGURE 18.6 LC filter L section.

output of the filter (V_{out}) is based only on the impedance Z_1 and Z_2; we can rewrite the voltage ratio transfer function $H(j\omega)$ as

$$H(j\omega)=\frac{z_2(\omega)}{z_2(\omega)+z_1(\omega)}\Rightarrow\frac{\left[\dfrac{1}{j\omega C}\right]}{\left[\dfrac{1}{j\omega C}\right]+\left[j\omega L+r_1\right]} \tag{18.12}$$

This simplifies to

$$H(j\omega)=\frac{1}{j\omega r_1C-\omega^2LC+1}\Rightarrow\frac{1}{(1-\omega^2LC)+j\omega Cr_1} \tag{18.13}$$

Resonance occurs at a frequency ω_o, where $\omega_o^2LC=1$ or $\omega_o^2=LC^{-1}$, and the Q factor of the LC structure is $Q=1/r_1\sqrt{L/C}$.

These are illustrated for the purposes of simplicity and are defined through analysis in section 18.6. The transfer function of Figure 18.6 may be described in standard form, which shows equivalence to the classical characteristic equation for a double-pole system.

$$H(j\omega)=\frac{1}{1-\left(\dfrac{\omega}{\omega_o}\right)^2+j\omega Cr_1}\rightarrow\frac{1}{\omega_o^2-\omega^2+j\omega\omega_o^2Cr_1}\triangleq\frac{\omega_o^2}{-\omega^2+j\omega\dfrac{\omega_o}{Q_o}+\omega_o^2} \tag{18.14}$$

We can see from equation (18.14) for $H(j\omega)$ that the denominator is a quadratic polynomial and has equivalence to the classical form (characteristic equation) that is used to describe second-order systems. In our case, it represents the dynamic behavior of a two-pole *LC* low-pass filter.

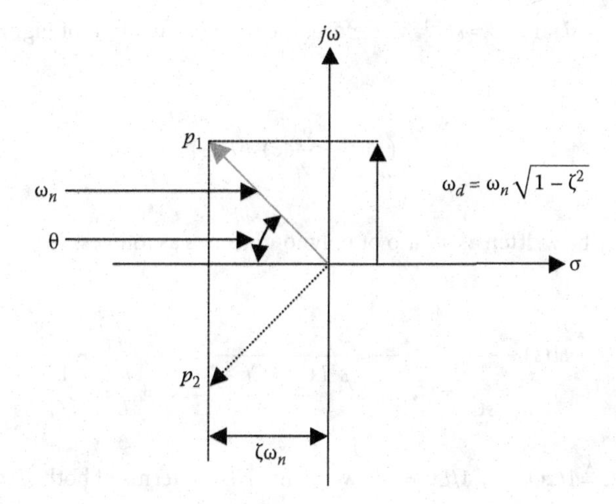

FIGURE 18.7 Complex S plane.

$$H(j\omega) = \frac{\omega_o^2}{-\omega^2 + 2\zeta\omega_o + \omega_o^2} \Rightarrow \frac{1}{1 + j2\zeta\left(\frac{\omega}{\omega_o}\right) - \left(\frac{\omega}{\omega_o}\right)^2} \tag{18.15}$$

The term zeta (ζ) in the denominator is used to describe the gain at ω_o, where $Q_o = (2\zeta)^{-1}$. When $\zeta < 1$, the roots of the quadratic p_1,p_2 may be defined as

$$-\omega^2 + 2\zeta\omega_o + \omega_o^2 = (s - p_1)(s - p_2) \tag{18.16}$$

$$p_1, p_2 = -\zeta\omega_o \pm j\omega_o\sqrt{1-\zeta^2} = -\zeta\omega_o \pm j\omega_d \tag{18.17}$$

where $\omega_d = \omega_o\sqrt{1-\zeta^2}$ is known as the damped natural frequency of the filter and provides both the coordinates and locus of the complex conjugate roots within the complex plane (Figure 18.7). This is very important when we consider the stability of the filter and is further discussed in section 18.7.

18.5 Transfer Function Analysis—*H*(*s*)

The transfer function *H*(*s*) is the frequency-domain description of a linear time-invariant system and is necessary for both analysis and synthesis in this domain. Analysis of circuit transfer functions is greatly simplified when the Laplace operator (*s*) is employed as a complex variable, where the complex frequency *s* replaces *j*ω, thus $s = \sigma + j\omega$ and provides the complex conjugate pole positions for p_1,p_2 in terms of both real and imaginary components.

We must also note that each term in *H*(*s*) has the dimension of ohms. If we apply the following terms in (*s*) for the generalized impedances in both the inductor and capacitor, we can write $L = sL$ and $C = sC^{-1}$. Therefore, the transfer function of Figure 18.6 may be written as follows:

$$\frac{v_o}{v_i}(j\omega) \to \frac{v_o}{v_i}(s) = H(s)$$

where $H(s)$ may be written as a ratio of polynomials in s as follows:

$$H(s) = \frac{\frac{1}{sC}}{\frac{1}{sC_1} + sL + r_1} = \frac{1}{s^2LC + sCr_1 + 1} \triangleq \frac{\frac{1}{LC}}{s^2 + \frac{r_1}{L}s + \frac{1}{LC}} \qquad (18.18)$$

Let $r_1/L = 2\zeta \triangleq 1/2Q_o$ and $1/LC = \omega_o^2$; we define $H(s)$ in terms of both ω_o and ζ in equation (18.19).

$$H(s) = \frac{1}{1 + 2\zeta\left(\frac{s}{\omega_o}\right) + \left(\frac{s}{\omega_o}\right)^2} \qquad (18.19)$$

Then with $2\zeta = \omega_o/Q_o \Rightarrow Q_o = \omega_o L/r_1$, we can write the solution for $H(s)$ in classical form, as shown in equation (18.20).

$$H(s) \triangleq \frac{\omega_o^2}{s^2 + \frac{\omega_o}{Q_o}s + \omega_o^2} \qquad \frac{\omega_o^2}{s^2 + 2\zeta\omega_o s + \omega_o^2} \qquad (18.20)$$

For the purposes of clarification, the LC filter of Figure 18.6 is a two-pole structure with complex poles: $p_1, p_2 = -\zeta\omega_o \pm j\omega_o\sqrt{1-\zeta^2}$. The filter response can be described in terms of the pole-Q frequency (ω_o) and either the pole-Q factor, (Q_o) or the damping coefficient zeta (ζ). It is important to note that the transfer function $H(s)$ above does not include the interaction with both the source and load impedance. This is presented in the next section.

18.6 Coefficient-Matching Technique

Coefficient matching is a process of realizing a network transfer function for a particular frequency-magnitude response and in defining suitable scaling terms for both L and C in order to meet that response. This technique is really only useful for simple structures, as the derivation of transfer functions becomes more complex as the order of the filter increases. Generally, for EMI filter design, we really do not care about meeting a specific frequency-magnitude response. We do, however, want to minimize the circuit-Q so that the filter is suitably damped for both input transient behavior and potential interaction with the incremental negative load resistance, as seen in PWM converters. The coefficient-matching technique is really more applicable to passive filters that require impedance-matching properties for maximum power transfer and low reflection coefficient. From an EMI standpoint, the source and load will almost certainly be in an unmatched condition throughout the frequency bands of interest. This implies that the characteristic impedance of the filter at F_c will not match the source and load impedances and, therefore, the DC gain of the filter will change over frequency. Our fundamental need is to ensure that we have sufficient insertion loss over the frequency range to meet the EMC requirements whilst also ensuring that the filter Q ≤ 1, so eliminating oscillatory behavior.

For general second-order *LC* structures, this approach is a simple solution toward realizing a scaling factor for both *L* and *C*. This also ensures that the derivation of scaling terms is defined to minimize the natural Q of the circuit structure (inductive and capacitive reactances equal). Once a filter structure is in place and has actual values for *L* and C, additional dQ damping in the form of either shunt RC, or series LR may be added to further control the circuit-Q.

The circuit of Figure 18.8 is second order and is terminated at both R source (R_S) and R load (R_L). Perform transfer function analysis to yield *H*(*s*) using ratio of polynomials in *s*.

$$\frac{v_o}{v_i}(s) = H(s) = \frac{R_L \parallel z_2}{(R_L \parallel z_2) + z_1 + R_S} \tag{18.21}$$

$$H(s) = \left[\frac{1}{\dfrac{R_L}{1+sCR_L} + sL + R_S} \left(\frac{R_L}{1+sCR_L} \right) \right] = \frac{\dfrac{1}{LC}}{s^2 + \dfrac{L + CR_L R_S}{LCR_L} s + \dfrac{R_L + R_S}{LCR_L}} \tag{18.22}$$

We can see that the expression is of the generalized quadratic form

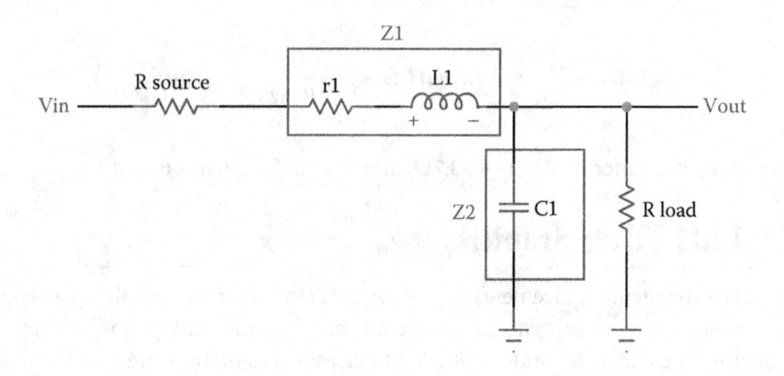

FIGURE 18.8 LC section with resistor terminations.

$$H(s) \triangleq \frac{\omega_o^2}{s^2 + 2\zeta\omega_o s + \omega_o^2} \tag{18.23}$$

Therefore, we can perform a coefficient-matching process to the equivalent frequency-magnitude response. For the purposes of example, we shall match coefficients for a Butterworth response where the poles of *H*(*s*) are $p_1, p_2 = -0.707 \pm j0.707$. In doing so, we normalize the transfer function for *H*(*s*) above and this makes $R_S = R_L = 1$, which simplifies the analysis greatly. We now rewrite *H*(*s*) in normalized form as follows and compare it to the Butterworth normalized quadratic.

$$H(s) = \frac{\frac{1}{LC}}{s^2 + \frac{L+C}{LC}s + \frac{2}{LC}} \Rightarrow k\frac{1}{s^2 + \sqrt{2}s + 1} \tag{18.24}$$

From equations (18.22) and (18.24), we may make the following observations:

$$k = \frac{R_L}{R_S + R_L} \tag{18.25}$$

$$\frac{L+C}{LC} = \sqrt{2} \quad \frac{2}{LC} = 1 \quad k = \frac{1}{LC} \tag{18.26}$$

Therefore, the values of the normalized second-order Butterworth filter with $\omega_o = 1$ rad/s and $R_S = R_L = 1$ are $L = \sqrt{2}HC = \sqrt{2}F$, and the DC gain of the filter is $k = 0.5$ or −6 dB. For the purposes of example, we design a Butterworth filter with equal 50-ohm terminations for both R_S and R_L and with a pole-Q frequency of 6.28×10^3 rad/s. The values for *L* and *C* are as follows:

$$R_S = R_L = R$$

$$L = \frac{\sqrt{2}(R)}{6280} = 11.25mH, C = \frac{1}{(\sqrt{2}R)6280} = 2.25\mu F$$

The filter impedance is $\sqrt{L/C} = 70.7\Omega$ and has a DC gain of −6.02 dB.

18.7 EMI Filter Stability

To meet the stringent requirements placed on EMI emissions, almost all power converters and power conversion systems require an input EMI filter. Due to the complexity in modeling the converter and input EMI filter as a system, they are usually designed separately. Problems most often arise when these systems are integrated together, as they incur mismatch or overlap in impedances where the input impedance of the load subsystem interacts with the output impedance of the source subsystem, $|Z_{in}| < |Z_o|$, thereby creating interactions and system instability. However, in most practical circuits, it is not possible to achieve nonminimal interaction. If Z_o and Z_{in} intersect, then the impedance overlap must be analyzed for stability.

EMI filters themselves are often responsible for many EMC test failures due to a range of issues. These include

- Impedance mismatch between adjacent sections within EMI filter structure
- Mismatch between input impedance (Z_{in}) of loaded system and output impedance (Z_o) of the source system
- Inherent high-Q source circuit with poor stability margin

Chapter 19 provides discussion and examples for stabilizing filter structures for pole-Q resonance and in ensuring that the filter output impedance is defined to ensure no overlap with the negative resistance of the load.

19

Filter Design Techniques and Design Examples

There are numerous techniques for designing filters. Some are based upon formal methods using numerical analysis; many others are based upon individual process, rules of thumb, and trial and error. Typically, EMI filters are used to reduce conducted emissions to an acceptable level so that a given test specification may be met. EMI filters may also be used to limit inrush current and suppress voltage transients caused by lightning and line transients. The specifications for the allowable interference are generally driven by the power circuit specification. The most common specifications include MIL-STD-461 for military applications, while CISPR and DO-160 are used for commercial applications. Many other EMI specifications also exist. This chapter provides technical discussion and analysis, including techniques that may be employed in the design of EMI filters that reduce conducted interference. The design of the input filter is slightly more critical when the power circuit topology is a regulated switching circuit such as a PWM power supply or a motor controller, etc., rather than a linear circuit. This is primarily due to the incremental negative input resistance that is a phenomenon of a switching circuit. This factor alone drives specific needs for the filter to ensure system stability and is presented in this chapter. The EMI filter design examples in this chapter assume that the load is a PWM power converter and not a linear load regulator.

19.1 Filter Design Requirements

The design of any EMI filter must start with a formal requirements specification such as MIL-STD-461, DO-160, or some other requirement such as CISPR. These all essentially define the limit levels for both conducted emissions (CE) and radiated emissions (RE) over a band of frequencies. Once this has been established, the filter design may be considered at a high level in order to capture accurate design drivers. These are key factors that will impact the filter in terms of performance, characteristics of source and load, filter size and weight, cost, etc. With these in place, a filter design specification can be developed, which ultimately will drive the design process. Before a filter is developed, there are other key factors and system constraints that will dictate the final solution. These are

- Form factor limitations and mechanical constraints
- Mounting types/packaging needs
- Environmental conditions (temperature, shock, vibration, moisture)
- Electrical characteristics (voltage, current, capacitance, insertion loss)

For the purposes of clarity, we shall focus on the circuit design aspects of an EMI filter and in meeting the requirements for conducted emissions and conducted susceptibility.

19.2 Design Techniques

There are several methods to use in designing an EMI filter, and they all have their own merits and weaknesses. Needless to say, the goal of any EMI filter design effort is to ensure that both conducted and radiated emissions testing is successful, and that the requirements are met. Typically, the design approach for an EMI filter will fall into one of the following:

1. *Intermediate testing.* Design the filter based upon both intermediate testing of the equipment that needs the filter, such as a PWM power converter, etc. The testing will cover conducted emissions for CM and DM loss assessment.
2. *Previous experience.* Use an existing filter or design with a known performance index and hope for good test results. The approach may include using a known filter structure with some manipulation of component values.
3. *Analysis, synthesis, and simulation.* Design the filter based upon analysis and simulation techniques.

These different approaches are discussed in the following subsections.

19.2.1 Intermediate CE Testing

This section covers intermediate testing (item 1 in the previous list) and assumes prefilter design test and assessment of harmonic composition.

1. Test power converter for conducted emissions without the EMI filter
2. Define insertion loss needs to include safety margin
3. Design filter, implement, and test
4. Optimize filter for insertion loss and stability

This approach is almost certainly the lowest-risk method, as the filter is designed against actual measured data, which enables accurate estimates for insertion loss. In reality, the procedure of pre-EMI filter testing is not used too much due to the need to partition a product-design effort and conduct expensive CE (conducted emissions) testing mid program. Many programs simply do not have the time to do this, as they are often struggling to meet customer schedules and milestones. If time permits, this approach is invaluable for military and commercial applications, where the cost of qualification failure is likely to impact product delivery.

19.2.2 Previous Experience—Similar Application

Where an application is duplicated, or has similarities in both power conversion architecture and voltage, current, etc., the use of a "standard EMI filter" that has previously passed testing is often used where the risk of failure is considered comparatively small. Is this really a true statement? This approach is cost effective and relatively low risk so long as the systems using the filters are very closely compatible. It must be said that there cannot be a perfect plug-and-play duplication of an EMI filter, as each PCB, each PWM converter, and each power system has its own set of peculiarities and associated parasitic uncertainties. If the filter has built-in margin and may also be test-setup tuned for insertion loss or stability, then this may very well be a good approach for a production or volume product.

19.2.3 Analysis, Synthesis, and Simulation

The concept of developing a filter based solely upon academic rigor and simulation is often daunting. Nevertheless, this is actually a very robust method if the analysis and simulation are done correctly. This approach is most often carried out during the design phase of the power conversion system and may also be done to mitigate the risk of the design phase for weight, form factor, performance, etc. It should also be mentioned that the design of a common-mode filter is based upon a completely different data set compared to a differential-mode filter; therefore, these two filters constitute two very different designs even though they sit together in cascade. The question arises, "What is the design process for this approach?"

To design an EMI filter effectively, knowledge of a time-varying signal is required so that the frequency spectrum (FFT) may be obtained and analyzed for differential-mode power spectrum estimates. Much of the success to this approach largely depends upon the quality of the simulation, analysis, and derivation of design data. In many cases, a simple equivalent model of a power converter topology under PWM control is sufficient. The objective of the simulation is the evaluation of the harmonic content through Fourier decomposition and analysis of the current signature. This may be a flyback operating in either continuous or discontinuous mode, or a phase-shifted bridge, or even a brushless DC motor controller. In most cases, we are able to reconstruct the current signature (be it, drain current, DC link current, etc.), and this may be used in our simulation. The FFT decomposition will yield valuable data and allow the designer to make critical assessment of where insertion loss is needed relative to frequency in order to meet a unique specification.

This approach is relatively robust for differential-mode loss; however, in the case of common mode, parasitic influence, stray capacitance in PCB layout, wiring, etc., all have a major influence in determining common-mode power density. In this case, it is recommended that the EMI filter either be designed very defensively or a CM test should be performed prior to the CM filter design. Once again, this is often not an option; experience and a robust design solution are necessary for a right first-time design. Once an accurate harmonic assessment has been made, and compared to a defined limit level such as CISPR, DO-160, or MIL-STD-461, etc., the loss required to attenuate the highest harmonic, including the –3-dB pole-Q frequency, may be calculated. From there, the filter structure and topology, along with suitable values for *L* and *C* including dQ networks, may be defined and selected.

19.3 Filter Design Summary

We shall present two examples of a filter design process. The first will follow a design flow with both analysis and simulation for a two-pole filter. The second example will expand on the two-pole design and develop a four-pole filter that is optimized for component selection, selection of filter impedance, stability, and insertion loss.

19.3.1 Predesign Objectives

Before we are able to define an EMI filter solution, it is necessary to have a good understanding of the harmonic composition, or content that reflects the switching current waveform. This is the unique signature that ultimately defines the frequency bands where the threat exists. More importantly, this data will drive the filter design in terms of insertion loss versus frequency. To realize this, we must create a simplistic but accurate model (e.g., Spice, etc.) and perform an FFT sweep to capture the spectral composition. The amplitude component of the Fourier spectrum or power density spectrum describes how much energy is contained in each frequency component within the signature FFT. Assuming that the FFT is based upon a good approximation of this current signature along with any dominant parasitic effects within our simulation model, we are able to overlay the specific dB amplitude limit requirements (dBμV or dBμA) onto the FFT data and verify the insertion loss needs versus frequency. If the precise current waveform is not known, then we may estimate this with reasonable accuracy in order to drive a baseline solution for the filter. It is important to remember that this is a process of simulation and, as such, it will never be perfect; therefore, accurate approximations with safety margin are all that can be achieved.

19.3.2 Define Design Flow

The filter design flow includes a set of design stages, some of which may be omitted or modified; however, for clarity, we present a description of the various stages as follows:

1. Verify EMC requirement.
2. Specify input voltage and current handling requirement, including any transient voltage conditions such as DO-160 or lightning.
3. Define inrush protection circuit needs if applicable, as this will drive specific filter needs.
4. Define PWM converter negative resistance. To recap discussion in chapter 18:
   A PWM-based power converter is designed to hold its output voltage constant no matter how its input voltage varies. Given a constant load current, the power drawn from the input supply is, therefore, also constant. If the input voltage increases by some factor, the input current will decrease by this same

factor to keep the power level constant. In incremental terms, a positive incremental change in the input voltage results in a negative incremental change in the input current, causing the converter input impedance to look like a negative resistor at its input terminals. Therefore, the instantaneous value of the input impedance is positive, but the incremental or dynamic impedance, or resistance, is negative. The value of this negative resistance depends on the operating point of the converter according to

$$R_n = -\left|\frac{V_{in}}{I_{in}}\right| \tag{19.1}$$

where R_n is defined as the incremental negative resistance of the power converter.

5. Based on the negative resistance, define EMI filter output impedance limit such that $|Z_O << R_n|$. We should also consider that R_n is smallest when at full load and at low line input voltage.
6. Define current signature or estimate current waveform.
7. Option (A): Simulate PWM topology in Spice.
   - Measure differential-mode current within model
   - Perform FFT and define magnitude (dBV, dBμV, dBμA) of fundamental harmonic
8. Option (B): Based upon estimation of current waveform, approximate magnitude of fundamental term in FFT using numerical methods. This can be complex and is not advised.
9. Define insertion loss requirements for EMI filter.
10. Define −3-dB pole-Q frequency of filter (for second-order filters). Four-pole filters will use two different pole-Q frequencies to separate the two resonances. Typically, we would specify, as a minimum, an octave for pole-Q separation with four-pole structures.
11. Define filter structure, number of poles, and method(s) of stabilizing the filter.
12. Calculate component values for L and C ensuring that for filters with four poles or more, the impedance of each section is $<R_{in}$.
13. Verify that the filter output impedance $Z_O << R_n$.
14. Define the filter stability factor (Q) for an input step response. We must remember that we have two causes of instability, the first being attributed to the capacitive and inductive reactance being equal at the pole-Q frequency. This will lead to resonance where the peak amplitude of the resonance is determined by the filter Q, or Q factor: $Q = \omega L/r_{dc}$. If Q is too high, a small disturbance may lead to an oscillatory response. The second cause of instability may also exist due to impedance interaction between the EMI filter output impedance and PWM converter incremental negative resistance. This is of concern and will drive the careful selection for filter output impedance.
15. Add dQ damping as needed within the filter structure to compensate for a higher Q and to modify (reduce) the filter output impedance if necessary.

19.4 EMI Filter Design Example

For the design example, the power converter is a switch-mode power supply of type flyback and is operating in discontinuous mode at a modulation frequency of 150 kHz. Input voltage is isolated +28 V nominal with a range of 18 to 32 V DC with an output power of 100 W. This implies that the EMI filter input ground is not connected to a local ground or chassis ground, but at the source. Therefore, common-mode loss is also required for this filter.

Power converter efficiency (η) is approximately 90%. The conducted emissions requirement places maximum current amplitude of 2 mA to be reflected back to the source at 150 kHz. This is equivalent to 66 dBμV loss and meets the requirements for CISPR 22 class A.

19.4.1 Design Process

Based on the data supplied for the example in Table 19.1, we shall walk through the design process and provide discussion and derivation as follows:

Define negative resistance as follows:

$$P_{out} = P_{in}\eta \Rightarrow I_{in} = \frac{P_{in}}{V_{in}} \tag{19.2}$$

$$R_{in} = \frac{V_{in}^2}{P_{in}} \rightarrow \frac{V_{in}^2\eta}{P_{out}} = \frac{18^2(0.9)}{100} = 2.9\,\text{ohms} \tag{19.3}$$

This suggests that the filter output impedance Z_O will be ≤2.9 ohms to ensure system stability.

19.4.2 Define Peak Harmonic Amplitude

Our main goal here is the approximation of the insertion loss needed at 150 kHz and beyond (150 kHz is the start of sweep range for CISPR 22), and we have two options available.

The first option is to develop an accurate Spice switching model of our PWM converter to replicate the current signature. The model should employ an LISN on both the source and return inputs. Then we simply perform an FFT on the differential-mode current (at the LISN 50-ohm outputs) within the circuit to establish the peak harmonic amplitude

TABLE 19.1 Limits for CISPR 22 Conducted Emissions

Frequency Range (MHz)	Average Limits dB(μV)
0.15–0.5	66
0.5–5.0	60
5.0–50	60

of the fundamental. Once a plot of the harmonic spectrum exists, we can either draw in or overlay the desired amplitude limits in dB versus frequency. By inspection of the graph data, we are then able to define the attenuation required at the fundamental harmonic to ensure that it is below the limit with, as a minimum, 6-dB error margin. Once we know how much loss is required (may use dBV or dBμV, etc.) at 150 kHz, we are then able to define the corner frequency, or pole-Q frequency, of the filter based upon numerical derivation using an *n*-pole structure. This derivation will assume, for the purposes of example, a two-pole structure, and allow us to calculate the −3-dB pole-Q frequency based upon the loss at 150 kHz. If the −3-dB frequency is too low to be practicable, we may then move to a four-pole structure. Once we have a solution that looks feasible, we may continue in terms of establishing practical component values and in stabilizing the filter Q.

The second option assumes that we don't have the means to develop a Spice circuit model, for whatever reason. If this is the case, we may want to, as a minimum, find a simple approximation for the peak harmonic amplitude of the fundamental so that we can perform a similar process. To do this, we can derive an approximation as follows.

Firstly, Fourier series may be used to represent periodic functions as a linear combination of both sine and cosine functions. So, if $f(t)$ is a periodic function of period T, then its Fourier series can be given by

$$f(t) = \frac{a_0}{2} + \sum_{n=1}^{\infty}\left[a_n \cos\frac{2\pi(n)t}{T} + b_n \sin\frac{2\pi(n)t}{T}\right] \tag{19.4}$$

where $n = 1,2,3,\ldots$, and T is the period of that function. The Fourier coefficients are called a_n, b_n and are given by

$$a_0 = \frac{2}{T}\int_0^T f(t)dt \tag{19.5}$$

If we consider both a_n, b_n, we can write

$$a_n = \frac{2}{T}\int_0^T f(t)\left(\cos\frac{2\pi(n)t}{T}\right)dt \tag{19.6}$$

$$b_n = \frac{2}{T}\int_0^T f(t)\left(\sin\frac{2\pi(n)t}{T}\right)dt \tag{19.7}$$

In the simplest case, a periodic function $f(t)$ of a 50% duty cycle square wave is defined as

$$f(t) = \begin{cases} (1) \rightarrow 0 \le t < T/2 \\ (-1) \rightarrow T/2 \le t < T \end{cases} \tag{19.8}$$

The solution to this periodic function for a_0 may be shown as

$$a_0 = \frac{2}{T}\int_0^{T/2}(1)dt + \frac{2}{T}\int_{T/2}^{T}(-1)dt \tag{19.9}$$

If we substitute for equations (19.6) and (19.7) into equation (19.8), $a_0 = 0$, $a_n = 0$ where

$$b_n = \frac{2}{(n)\pi}1 - \cos[(n)\pi] \tag{19.10}$$

the Fourier series is as follows:

$$f(t) = \sum_{n=1}^{\infty}\frac{2}{(n)\pi}[1-(-1)^n]\sin\frac{2\pi(n)t}{T} \tag{19.11}$$

If we were to look at the peak amplitude of the fundamental harmonic, we could see that using equation (19.10), the amplitude factor, is equivalent to $2/[(n)\pi] = 0.636$ of the peak pulse amplitude, and 0.212 of the peak pulse amplitude for the third harmonic. This is a very conservative approximation and is applicable to a rectangular periodic waveform. For a triangular waveform, or indeed a waveform that represents the flyback drain current for discontinuous conduction mode, the net contribution would be somewhat smaller, and we would obviously need to modify the expression for $f(t)$ in equation (19.8) to capture this.

19.4.3 Define Harmonic Current

For the purposes of example, we assume that simulation data is not available. We also assume a square-wave approximation at 50% duty cycle; therefore, the average input current may be found as follows:

$$I_{av} = \frac{P_{out}}{V_{in}\eta} = \frac{100}{18(0.9)} = 6.17A \tag{19.12}$$

The peak amplitude will be twice this value for a 50% duty cycle or 12.3 A. The fundamental peak amplitude is, therefore, defined using equation (19.10). This is also shown in Figure 19.1.

$$\hat{I}_p^F = \frac{2}{\pi}(I_{pk}) = 7.85\text{ A} \tag{19.13}$$

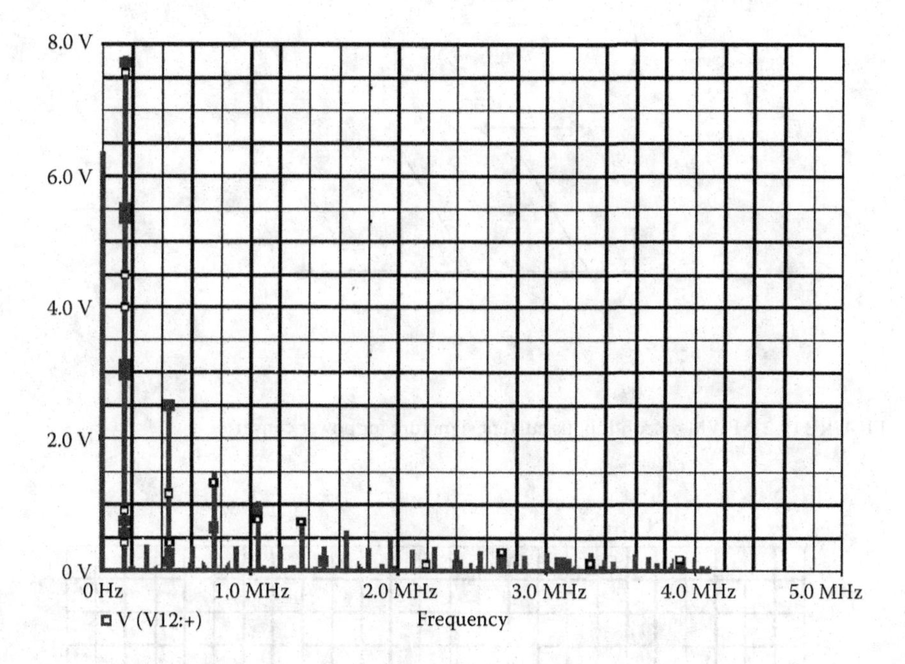

FIGURE 19.1 FFT composition for 150-kHz square wave.

From this value, we are able to realize the attenuation (α) required at the fundamental harmonic. Dividing the fundamental peak current by the 2-mA limit, this is simply

$$\alpha = 20\log_{10}\left(\frac{0.002}{7.85}\right) \cong -72\,\text{dB} \tag{19.14}$$

or the attenuation factor A is

$$A = \frac{7.85}{0.002} = 3910 \tag{19.15}$$

This is also the same as writing

$$\alpha = 20\log_{10}\left[\frac{f_{-3\text{dB}}}{F_{\text{pwm}}}\right]^2 \triangleq 40\log_{10}\left[\frac{f_{-3\text{dB}}}{F_{\text{pwm}}}\right] \Rightarrow f_{-3\text{dB}} = \left[\frac{F_{\text{pwm}}}{10^{\frac{\alpha}{40}}}\right] \tag{19.16}$$

For a rectangular pulse, this would be a good approximation; however, a flyback converter in discontinuous conduction mode would lead to reduced fundamental harmonic amplitude due to the triangular drain current waveform, as shown in Figure 19.2. In this

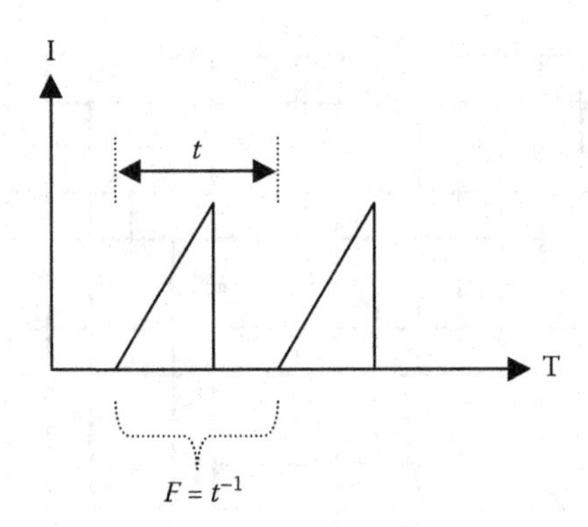

FIGURE 19.2 PWM discontinuous current signature for power converter.

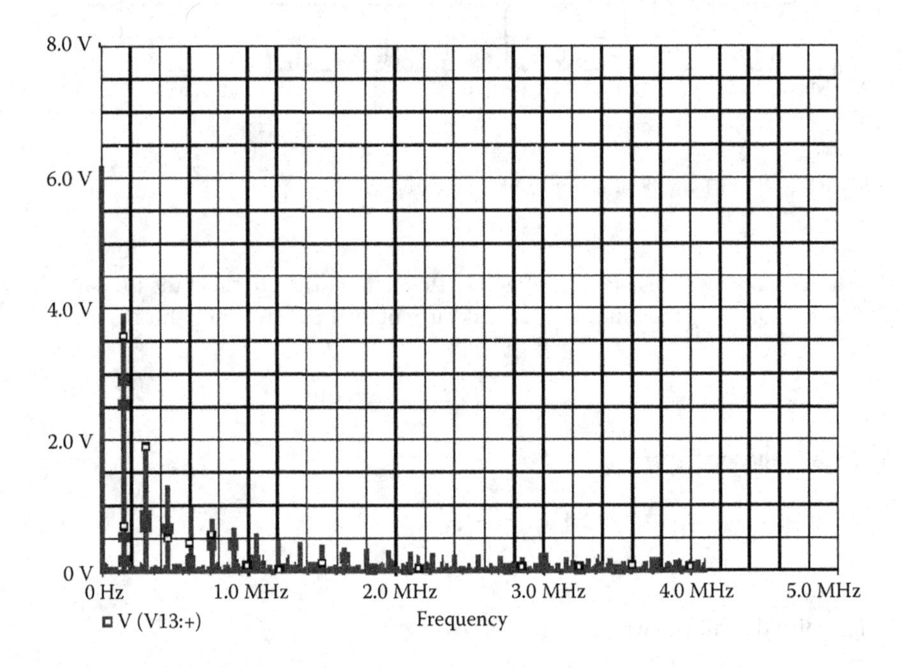

FIGURE 19.3 FFT composition for 150-kHz triangular wave.

case, the filter would not require such high levels of attenuation as shown in equation (19.14) at the fundamental harmonic.

Figure 19.3 shows the FFT harmonic amplitudes for a triangular wave and, as we can see, at approximately 4.0 V, the amplitude of the fundamental harmonic is almost half that for the square-wave approximation, or −6 dB down. This will also slightly reduce the insertion loss needs for the filter at 150 kHz. For the purposes of clarity, we shall assume that the filter attenuation is based upon the triangular waveform; therefore,

$$\alpha = 20\log_{10}\left(\frac{0.002}{4}\right) = -66\,\text{dB} \tag{19.17}$$

The actual filter loss will be −66 dBV or 54 dBμV at 150 kHz. We shall also add a −6-dB margin for uncertainty, which will bring the insertion loss requirement at 150 kHz to −72 dB.

19.4.4 Define Filter −3-dB Pole-Q Frequency for Differential Mode

For the purposes of example, the filter will be second order (design constraint) and, therefore, the frequency-magnitude slope equates to −40 dB/decade, which also drives the location of the pole-Q frequency as follows:

$$\alpha = 10\log_{10}\left[1+\frac{\omega_S}{\omega_C}\right]^{2n} \triangleq 20\log_{10}\left[1+\frac{\omega_S}{\omega_C}\right]^{n} \tag{19.18}$$

where ω_S,ω_C are the frequencies of interest. $\omega_S = 942 \times 10^3$ rad/s, or 150 kHz, and ω_C is the half-power pole-Q frequency. The order of the filter has been defined as part of our system requirements, so we already know that $n = 2$.

Equation (19.18) is actually based upon a maximally flat response function, or Butterworth approximation, where the amplitude response or magnitude is specified as

$$|H(j\omega)|^2 = \frac{H_0}{(1+\omega_S/\omega_C)^{2n}} \tag{19.19}$$

This is known as the nth-order Butterworth, or maximally flat low pass response, and was first suggested by Butterworth.* The term H_0 is a gain constant (DC) for the response and is set to unity for the purposes of this example. As we are dealing with, as a minimum, two-pole structures, the magnitude response of the Butterworth response is

$$|H(j\omega)| = \frac{1}{\sqrt{1+\omega^4}} \tag{19.20}$$

From equation (19.20), the minimum-phase two-pole transfer function is

$$H(s)\frac{1}{s^2+\sqrt{2}s+1} \triangleq \frac{1}{(s-p_1)(s-p_2)} \tag{19.21}$$

from which we may show that the normalized poles are at $p_1,p_2 = -0.707 \pm j0.707$.

* Named after British engineer Stephen Butterworth.

We can see from equation (19.19) that when $\omega_S = \omega_C$ and $H_0 = 1$

$$|H(j\omega)| = \sqrt{0.5} \quad (19.22)$$

This is the half- power frequency where the magnitude $H(j\omega) = 20\log_{10}\sqrt{0.5} = -3\,\text{dB}$.

For (n)-pole LC filter structures, this approach is certainly a valid solution and may very well be employed as a general approximation, even though EMI filters are not typically designed for a unique amplitude response. Interestingly, Butterworth is a unique case where $|Q| = |\zeta| = \sqrt{2}^{-1}$; therefore, the filter response is suitably damped, certainly so for an EMI filter. Again, EMI filter design is all about insertion loss, maintaining stability, and in meeting the EMC requirements. It is not about precise placement of poles, etc. From a practical viewpoint, the more one can do to achieve these goals, the better are the chances of meeting the EMC requirements. For our design example, (n) = 2.

Using equation (19.18), we can define a –3-dB pole-Q frequency of 2.414 kHz for the filter.

$$72\,\text{dB} = 20\log_{10}\left[1+\frac{\omega_S}{\omega_C}\right]^n \Rightarrow 3981 = \left[1+\frac{942}{\omega_C}\right]^2 \quad (19.23)$$

$$\omega_C = \frac{942}{\sqrt{3981}-1} \cong 15\,\text{KRads/s} = 2.4\,\text{kHz} \quad (19.24)$$

There is another way to define the pole-Q frequency if we know the following:

1. Number of poles (n)
2. Frequency of highest harmonic within limit (f_1)
3. Actual attenuation in dBV, or dBμV, at the frequency above

Based upon the filter design example, we know that our filter has two poles, which leads to a frequency-magnitude slope of –40 dB/decade. We also know that our frequency of interest is 150 kHz, and finally, we need a loss of –72 dB at 150 kHz. Using this data, and applying equation (19.16), we can say

$$-72 = -40\log_{10}\left(\frac{f_1}{fc}\right) \quad \text{or,} \quad fc = \left[\frac{150\,\text{kHz}}{10^{\frac{72}{40}}}\right] \cong 2.4\,\text{kHz}$$

where the –40 equates to the frequency-magnitude slope in dB/decade frequency for a two-pole filter, and the –72 is obviously the filter attenuation required at 150 kHz. For clarification, the attenuation with $\omega_C = 15 \times 10^3$ rad/s or 2.4 kHz is

$$\alpha = 20\log_{10}\left[1+\frac{942}{15}\right]^n \cong 72dB\Big|_{n=2} \quad (19.25)$$

19.4.5 Insertion Loss Validation

Insertion loss is a measure of the effectiveness of a filter in terms of attenuation. It is defined as the ratio of the two voltages: (V_1) across the circuit load without the filter and the voltage (V_2) across the load with the filter in circuit. Since insertion loss is dependent on the source and load impedance in which the filter is to be used, insertion loss measurements are defined for a matched 50-ohm system. Insertion loss may be defined as

$$IL_{dB} = 20\log_{10}\left[\frac{V_1}{V_2}\right] \tag{19.26}$$

Standard COTS filters, or those that are already manufactured, are almost always characterized between 50-ohm impedances. This is very unlikely to match the actual circuit impedance within a real EMI filter application. However, if the circuit impedances are known or can be closely approximated, it is possible to calculate the expected insertion loss from published 50-ohm values.

For example, with an LC low-pass L filter structure, the capacitor forms the transfer impedance Z_{12}, and this can be established in terms of the two-port network Z-parameters. The transfer impedance sits in parallel with the load impedance Z_L, and the actual attenuation of the filter in a system other than 50 ohms is not entirely deterministic unless the filter can be measured for attenuation within a 50-ohm system. If we had actual data for attenuation at 50 ohms, we can use (Figure 19.4) to define the transfer impedance. Defining the insertion loss for a system where both the source (Z_S) and load impedance (Z_L) are known (measurement or analysis) is achieved as follows:

$$\alpha_{dB} = 20\log_{10}\left[1 + \frac{Z_S Z_L}{Z_{12}(Z_S + Z_L)}\right] \tag{19.27}$$

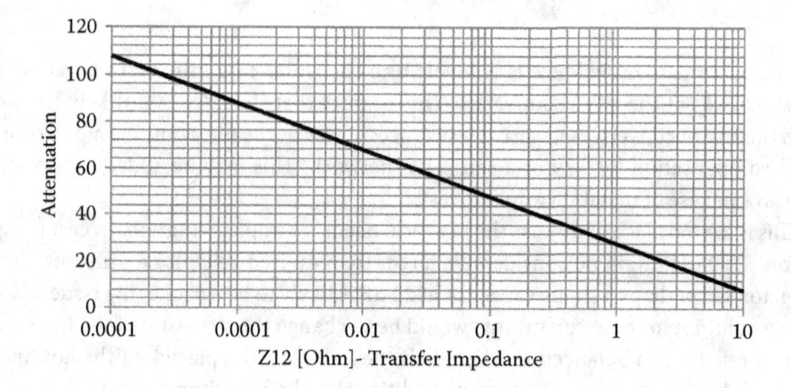

FIGURE 19.4 Transfer impedance for 50-ohm systems.

Of course, EMI filters are often designed for low-output impedance compared to that of the load, or to meet the needs of incremental negative resistance, and for that reason, a 50-ohm filter is often not practical.

19.4.6 Design Example Summary

Before components are selected for the design example, the filter design can be summarized as follows:

1. Filter order = 2
2. Pole-Q frequency = 2.414 kHz or 15 K Radians/sec
3. Filter loss at 150 kHz = –72 dBV
4. Filter output impedance $Z_O \leq 2.9$ ohms
5. Filter design impedance $R_d = R_{in}$

19.4.6.1 Define Component Values

$$C = \frac{1}{\omega_C R_d} = \frac{1}{2\pi(2414)2.9} \cong 22\mu F \tag{19.28}$$

$$L = \frac{R_d}{\omega_C} = \frac{2.9}{2\pi(2414)} = 190\mu H \tag{19.29}$$

19.4.6.2 Verify Pole-Q Frequency

$$f_c^{-3dB} = \frac{1}{2\pi\sqrt{LC}} = 2.414\,\text{kHz} \tag{19.30}$$

19.4.6.3 Define Characteristic Impedance of Filter

$$|Z_O| = \sqrt{\frac{L}{C}} = 2.9\,\text{ohms} \tag{19.31}$$

The characteristic impedance is equivalent to the worst-case minimum negative input resistance R_{in} of the PWM converter. This implies that the filter is unlikely to interact with the PWM converters negative resistance; however, the filter output impedance may need compensation by way of a shunt RC network. This may be added as a defensive measure in case of instability during test.

This is a good place to discuss the practical needs for implementation. In certain applications, 190 μH might be considered a larger than desired value for a differential-mode inductor. Or perhaps a single capacitor at 22 μF is likely to be a packaging issue due to its size. A solution to these constraints would be to change the structure of the filter from a single-ended L to a balanced L, where half the inductance is placed on the hot line and the other half is placed on the return. Additionally, the capacitor may be split to form a π filter. The filter may be further optimized for smaller values and smaller components

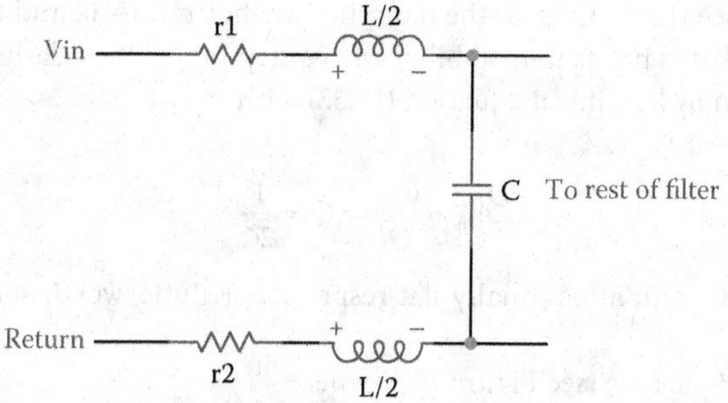

FIGURE 19.5 Balanced L filter structure

if the design were based upon four poles, or two LC structures in cascade. This is discussed later in the chapter. For now, we shall assume a balanced L structure as shown in Figure 19.5 above where the total inductance is halved and each inductor is approximately 95 μH. The terms for r_1 and r_2 represent the individual ESR for each inductor.

19.4.6.4 Stabilize the Filter

Passive LC filters have inherent high Q characteristics due to the reactance of both L and C being equal at the resonant frequency. Without damping, this leads to highly selective response at the pole-Q frequency, thereby creating system instability. If the filter-Q is too high, an increase in emissions can occur at the resonant frequency of the filter due to the increase in amplitude of the peak. At the resonant frequency, the filter will have a voltage gain such that the output voltage is Q times the input voltage, and this may create overvoltage issues with power conversion circuits. Input and output impedances are also affected by filter-Q. If Q is greater than 1, the input impedance will be lower than the characteristic impedance of the filter. Furthermore, the output impedance of the filter will be greater than the characteristic impedance. In both cases, the mismatch in impedance relative to the characteristic impedance is a factor of Q. These factors necessitate that the filter be suitably designed for optimal Q where $Q \le 1$. Optimization of an EMI filter refers to selection of the damping element such that the peak filter output impedance is minimized. There are two major drivers for stability: circuit-Q factor and interaction of the filter output impedance with the load. Before we look at the filter output impedance, we need to look at the filter structure itself and the circuit-Q that is attributed to the complex impedance $j\omega L$ versus the real resistance in the circuit, r_{DC}. This is the quality factor of the inductor, or

$$Q = \frac{j\omega L}{r_{DC}} \tag{19.32}$$

If we consider the characteristic equation for a second-order system in quadratic form, we have

$$H(s) = \frac{\omega_n^2}{s^2 + 2\zeta\omega_n s + \omega_n^2} \Leftrightarrow \frac{\omega_n^2}{s^2 + \left|\frac{\omega_n}{Q}\right| s + \omega_n^2} \tag{19.33}$$

We can clearly see that if Q >> 1, the damping coefficient $\zeta \rightarrow 0$, and this will lead to a filter structure that is prone to instability with a step input. The relationship between Q and ζ can be seen by looking at equation (19.33) with $\omega_n = 1$.

$$2\zeta\omega_n = \frac{\omega_n}{Q} \Rightarrow Q = \left.\frac{1}{2\zeta}\right|_{\omega_n=1} \tag{19.34}$$

As already mentioned, a maximally flat response, or Butterworth shows that Q = ζ = 0.707.

This is the unique case (see Figure 19.6) where

$$\zeta\omega_n = \omega_n\sqrt{1-\zeta^2} \tag{19.35}$$

The poles of the normalized denominator polynomial are located at

$$p_1, p_2 = -\sigma \pm j\omega = -0.707 \pm j0.707 \tag{19.36}$$

We can also state that the damping coefficient ζ can be found by

$$\zeta = \cos(\theta) = 0.707 \tag{19.37}$$

$$\theta = \left[\tan^{-1}\frac{j\omega}{\sigma}\right] = 45° \tag{19.38}$$

With a higher than desired Q factor, the filter will not become completely unstable. However, the oscillatory nature of the filter is likely to be unacceptable for practical use.

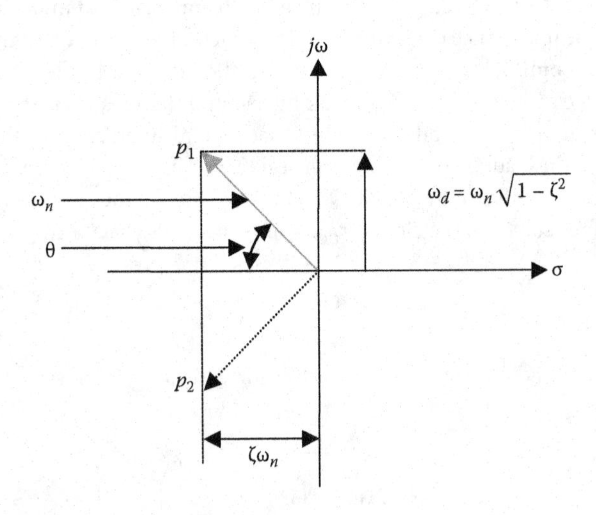

FIGURE 19.6 S-plane.

Typically, we want to aim for a filter Q of 1 or less. Figure 19.6 illustrates the use of the S-plane to show pole placement for the complex conjugate pole-pairs.

To summarize, adequate filter damping may be necessary to avoid destabilizing the feedback loops of DC-DC converters with duty cycle control. See references [1, 2].

19.4.6.5 RC Shunt dQ Damping

While this filter provides the proper impedance matching and the required attenuation, the impedance will be high at the resonant frequency of the filter. The only damping elements in the circuit are the equivalent series resistance (ESR) of both the inductor (r_L) and capacitor (r_C), both of which may amount to approximately 30 mΩ (estimated for our example). With $Q = j\omega L/R_{DC}$, we can see that at the pole-Q frequency, the inductive reactance of the 190 μH equates to $j\omega L = 2.9\ \Omega$; therefore, the resulting Q factor will be very high.

$$Q = \frac{j\omega L}{r_C + r_L} \Rightarrow \frac{1}{r_C + r_L}\sqrt{\frac{L}{C}} \cong 100 \tag{19.39}$$

It is normally necessary to provide damping of the LC filter to restrict the impedance of the filter at the resonant frequency. This is achieved by using a shunt series RC network that is connected in parallel to the filter capacitor. The value of the damping capacitor (C_d) is generally three to five times greater than that of the filter capacitor so as not to impact the pole-Q, or filter resonant frequency. The damping resistor (R_d) is simply equal to the characteristic impedance of the filter for optimal Q.

Figure 19.7 shows a balanced L filter structure with a shunt RC network. The angular frequency at which the damping network takes effect is

$$\omega_d = \frac{1}{R_d C_d} \tag{19.40}$$

This frequency is set below the filter pole-Q frequency, where R_d functions as an AC-coupled shunt resistor. Capacitor C_d blocks DC current and eliminates high power

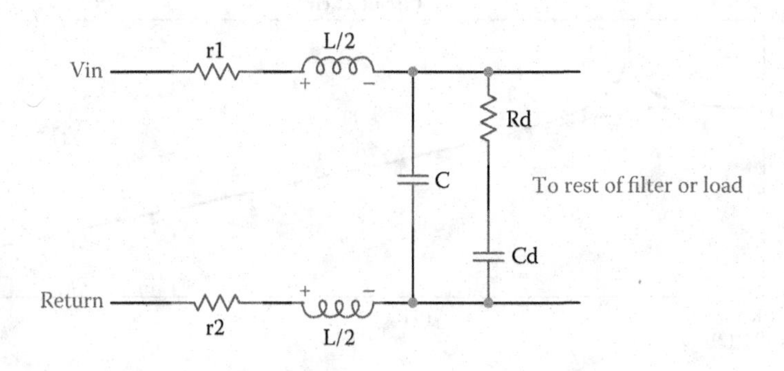

FIGURE 19.7 Balanced L filter structure with dQ RC shunt.

dissipation in R_d. For R_d to damp the filter, C_d must have an impedance magnitude that is sufficiently less than R_d at the pole-Q frequency (Figure 19.8). The filter high-frequency attenuation is not affected by C_d, and the high-frequency asymptote is the same for an equivalent filter without damping. See Figure 19.9.

There is a trade-off with the shunt RC approach, as the blocking capacitor will be significantly larger than the filter capacitor. For optimal Q, where Q ≤ 1, n is derived from the transfer function of the LC structure and equivalent load impedance.

$$Q_{opt} = \sqrt{\frac{(2+n)(4+3n)}{2n^2(4+n)}} \tag{19.41}$$

where n is the ratio of the two capacitor values

$$n = \frac{C_d}{C} \tag{19.42}$$

In the case of our design example, we would select the following components for the shunt dQ network: $R_d = R_{in} \approx 3$ ohms and $C_d = 4C|_{n=4}$. The damping capacitor C_d is 88 μF.

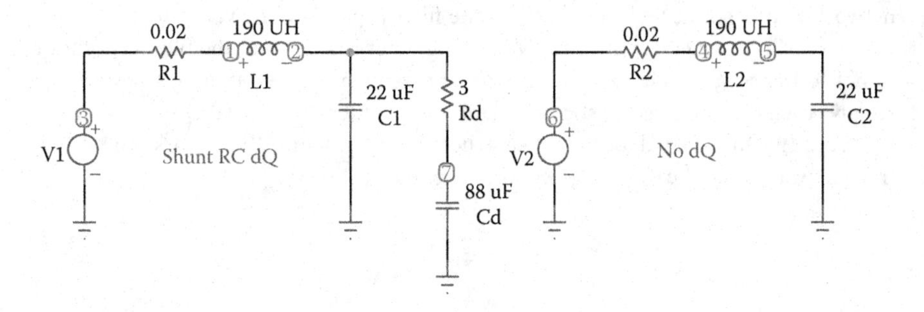

FIGURE 19.8 Shunt RC damping.

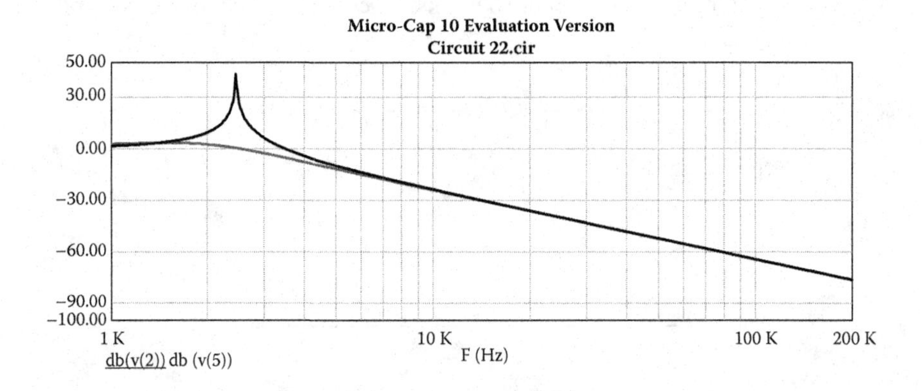

FIGURE 19.9 Frequency magnitude response for Figure 19.8.

The value of C_d is both large and almost certainly impractical for many applications where cost and form factor are major design drivers. Of course, trying to get away with a significantly lower value will degrade the high-frequency loss, and this may prove to be an issue. The large value of capacitor may also drive the use of series LR dQ damping.

19.4.6.6 Series LR dQ Damping

Figure 19.10 illustrates the placement of damping resistor r_d in parallel with the filter inductor L. Inductor L_d causes the filter to exhibit a two-pole attenuation characteristic at high frequency. To allow R_d to damp the filter, inductor L_d should have an impedance magnitude that is sufficiently smaller than R_d at the filter pole-Q frequency. To design the series dQ network for optimal Q where Q ≤ 1 (i.e., the choice of R that minimizes the peak output impedance, for a given choice of L_d), we can use the following expression

$$Q_{opt} = \sqrt{\frac{n(3+4n)(1+2n)}{2(1+4n)}} \tag{19.43}$$

where, n is the ratio of the filter inductance compared to the damping inductance

$$n = \frac{L_d}{L}$$

With this approach, inductor L_d can be physically much smaller than L. Because R_d is typically much greater than the DC resistance of L, essentially none of the DC output current flows through L_d. Furthermore, R_d may be realized as the equivalent series resistance of L_d at the filter pole-Q frequency. This approach offers a solution that eliminates the need for bulky capacitors. There is a disadvantage to this approach in that the high-frequency attenuation of the filter is degraded where the frequency-magnitude asymptote is modified from $1/\omega^2 LC$ to $1/\omega^2(L||L_d)C$. To get around any issues with

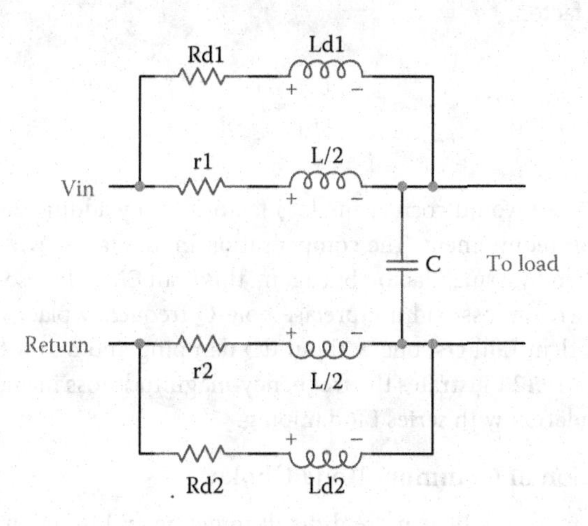

FIGURE 19.10 LC filter structure with RL series dQ network.

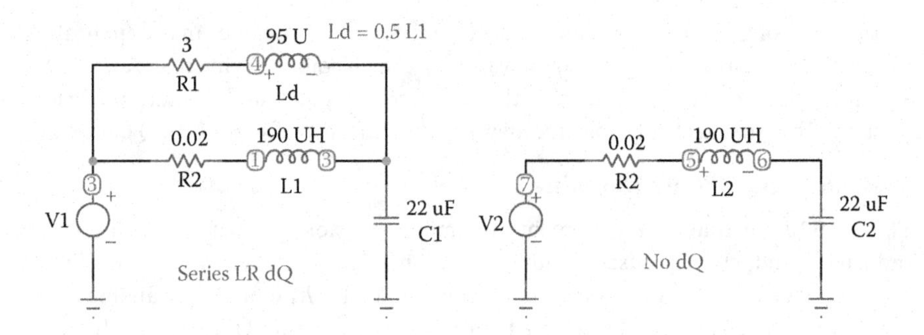

FIGURE 19.11 Series LR damping.

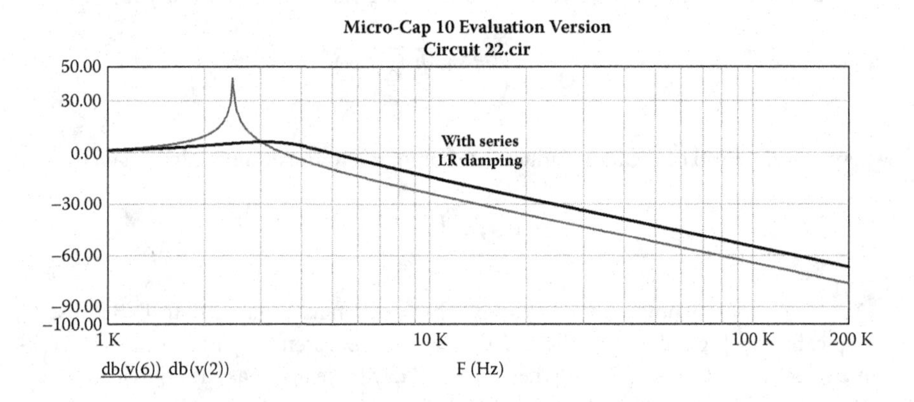

FIGURE 19.12 Frequency-magnitude response of Figure 19.11.

high-frequency attenuation, the filter can be design-corrected to compensate for frequency-magnitude degradation. The attenuation of the filter high-frequency asymptote is degraded by a factor

$$\frac{L}{L \| L_d} = 1 + \frac{1}{n} \tag{19.44}$$

So, for $n = 0.5$, we would correct the loss factor of 3 by adding $20 \log_{10}(3) = 9.5$ dB to the attenuation requirement. The compensation in loss factor will move the pole-Q frequency to the left by a small factor, but again, this is an EMI filter, so we really are concerned about insertion loss and not precise pole-Q frequency placement. Figure 19.11 shows two equivalent L filters, one without dQ damping and one with series LR compensation. Figure 19.12 illustrates the frequency-magnitude loss including the high-frequency dB degradation with series LR damping.

19.4.6.7 Addition of Common-Mode Choke

Common-mode noise is a little more difficult to define and is largely attributed to the capacitive displacements due to relatively high *dv*/*dt* action at a particular switching node. In the case of a flyback, for example, this would be the MOSFET drain. The switching action of power devices, such as a MOSFET, forces current to flow in the parasitic capacitive elements. This current is common to both source and power return and will flow to ground and return through these paths. Being parasitic dominant, the common-mode interference spectrum is predominantly higher in frequency compared to differential mode. The noise spectrum for common mode usually runs from the 10-kHz to 50-MHz range, and can appear anywhere between.

Predicting common mode is complex, and really, any stray parasitic capacitance effects will come into play if certain design constraints are not in place. These stray parasitic elements are also nonsymmetrical, and common-mode current distribution between source and return may not be equal, which gives rise to a further addition in differential-mode interference. In our experience, common-mode filter design must be defensive. Unless one is able to actually measure the common-mode noise in the final equipment design—and before EMI filtering is developed—use as much common-mode loss as is practicable. Managing common-mode interference also extends itself to the design of the power converter or the equipment that is generating the noise components. The most essential factors that need to be addressed are as follows:

1. Reduce the magnitude of the common-mode source by careful selection of *dv*/*dt* on power switching devices
2. Reduce parasitic capacitance in the circuit, in particular, within areas that surround the high-frequency switching power devices, power device to heat sink, primary to secondary of transformer, etc.
3. Reduce common-mode current with filtering, such as a DC link CM filter or LC output filter
4. Return CM current to its source through a small loop area, not an external ground

Grounding, filtering, isolation of power switching from control logic, minimizing parasitic capacitances, controlling power switch *dv*/*dt* rise times, etc., will all play a role in reducing common-mode noise.

Common-mode rejection is typically based upon shunt capacitors connected from both line and neutral to chassis ground. In addition, a common-mode choke is placed in series with the differential-mode inductors. This ensures that high frequencies are low-impedance shunted to chassis, and that the impedance presented to common-mode noise is high (Figure 19.13).

The basic parameters required for a common-mode filter are input current, impedance, and the frequency over which the loss is required. The design drivers for the filter are very straightforward, and are as follows: How much volume is available for the filter, and what restrictions, if any, are in place for capacitance to chassis ground (leakage current limit), etc.? In simple terms, this implies: how big can the core be, and how much Y capacitance can be used? Both of these will drive the design for high-frequency loss. The last design constraint is very important as shunt, or Y capacitors (C-CM), are restricted in size for certain applications in order to limit leakage current requirements. This is very typical with both aerospace and medical applications and will drive a higher value for the common-mode inductance.

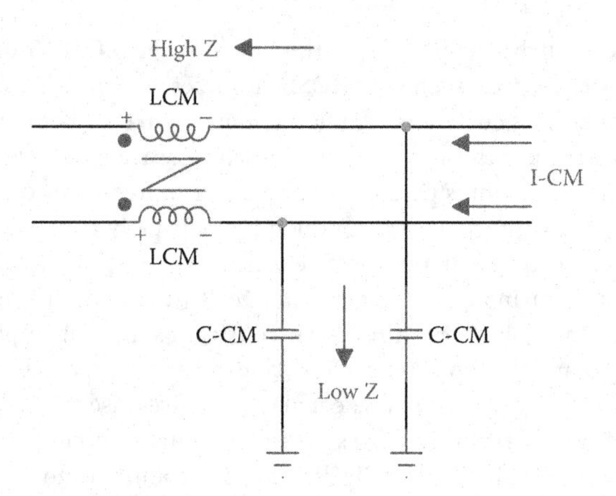

FIGURE 19.13 Balanced LC filter structure with common-mode components.

The first step in the design process is to look at the available volume within the filter enclosure. If the design has unique mechanical limitations or requirements, then the largest core should be selected so that it will meet the requirements when it is wound. The next step is to calculate the maximum number of turns that will fit on the core, and this may include bifilar windings for higher current. Once we have the maximum turns, we select an appropriate core material and define the inductance. The material selection is very important, as it will impact operating temperature and frequency of operation, or high-frequency loss. Once we have a figure for the permeability, we can make an assessment of the minimum inductance versus frequency and verify that we have sufficient inductance. If we don't, then we select another core material. If the calculated inductance is above the needs of the filter, we may select a smaller core and reduce the number of turns to make the design as practicable as possible. Common-mode filters operate over much wider frequency ranges than differential mode, and the core material performance at 100 kHz and beyond is something to consider. In general, high-permeability cores demonstrate degradation in performance at lower frequencies, where the inductance starts to roll off. This will reduce the effectiveness of the filter and may be of significance.

If we expand on Figure 19.10, we can add the common-mode filter to the differential-mode section, as shown in Figure 19.14. This is a balanced filter with additional distributed differential-mode capacitance, and includes series LR damping and an optional RC shunt at the filter output, which may be used to correct impedance.

19.4.6.8 Define Common-Mode Pole-Q Frequency

In some applications, the use of a common-mode choke is sufficient in terms of loss. For example, the largest core that will fit into the available space with good high-frequency performance is 3 mH. At a switching frequency of 150 kHz, the choke will present an impedance of $2\pi F_{pwm} L_{CM}$ = 2827 ohms. If we assume a 50-ohm load, this equates to a loss of –35 dB at the switching frequency. We know that when the

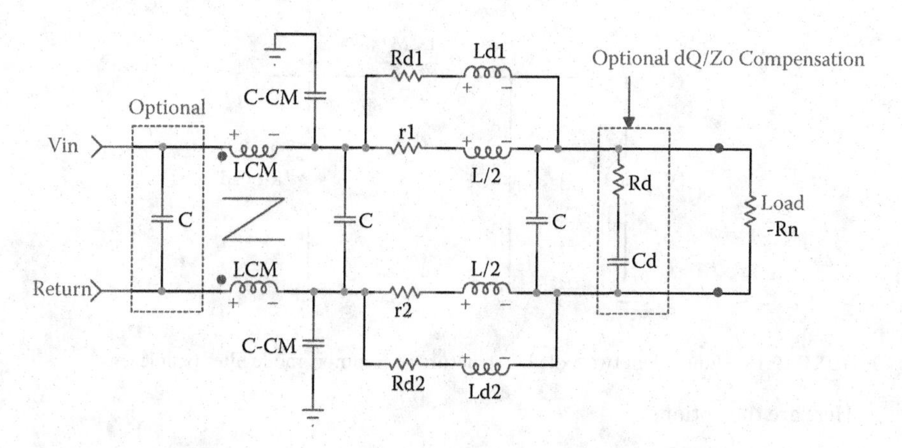

FIGURE 19.14 Balanced LC filter structure with common-mode components.

impedance of the choke is also 50 ohms, the loss will be −3 dB. Therefore, we can define this frequency

$$F_{-3\text{dB}} = \frac{50}{2\pi L} \cong 2.652 \text{ kHz} \tag{19.45}$$

Therefore, the attenuation of the filter at 2.652 kHz is −3 dBV, increasing at −20dB per decade. So, at the tenth harmonic of the switching frequency, or 1.5 MHz, the loss will be approximately

$$\alpha = 20\log_{10}\left[1 + \frac{1500}{2.652}\right]^{n} \cong 55.0 \text{ dB}\Bigg|_{n=1} \tag{19.46}$$

If increased common-mode loss in the lower frequency range is required, we can utilize the Y capacitors (C-CM), and these will create a two-pole filter structure, which has a loss of −40 dB per decade frequency. The implications of increased frequency-magnitude loss allow the inductance to be reduced so that a smaller core may be used, thereby permitting the use of larger Y capacitors if the requirements do not suggest otherwise. Either way, it is all about meeting the requirements, not just for EMC, but for cost, weight, and packaging.

In the case of the two-pole filter structure (Figure 19.15), a valid question might be, "Where do we place the −3-dB common-mode corner frequency?" We don't really know what the common-mode noise amplitude is going to be, or where in the frequency band might the problems occur. This is where a decision must be made, and one that will be decided by several factors. The first of these involves asking, "How much room is available for the common-mode filter?" Secondly, do we have a leakage current limit imposed on the design? If so, this will drive smaller Y capacitors. Smaller Y capacitors imply a larger choke, or inductance, value if the space is available.

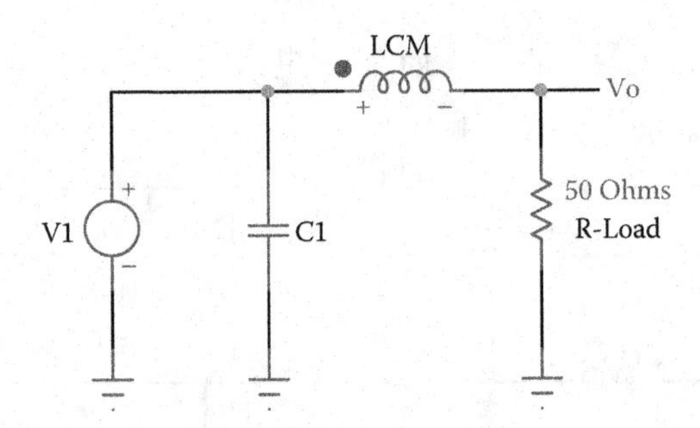

FIGURE 19.15 Equivalent two-pole LC structure of common-mode filter (single leg).

Here are the options:

1. Restriction on mechanical space and limit imposed on maximum capacitance to chassis. Use max allowable Y capacitors and largest inductance that will fit in the space available. The corner frequency will be a factor of these two constraints.
2. No restriction on mechanical space and no leakage current restriction. Use optimal inductance and capacitance based upon –3-dB corner frequency at ten times differential-mode –3-dB pole-Q frequency.

For the design example, we shall go with option 2.

Corner frequency f_{CM} of common-mode LC filter is defined as ten times that of equation (19.24), or 24.0 kHz. Next, we select a choke that will fit into the area and is practicable for manufacture while leaving sufficient room for other filter components, etc. This yields an inductance of 1.8 mH. We may now define the Y capacitor value as follows. From

$$f_{CM} = \frac{1}{2\pi\sqrt{LC_Y}} \tag{19.47}$$

$$C_Y = \frac{1}{4\pi^2 f_{CM}^2 L} \cong 0.022\ \mu\text{F} \tag{19.48}$$

Don't forget, this is not a critical frequency, nor does the capacitor need to be precisely 0.022 μF. This is merely a very good approximation, and values that are close to these will work very well.

As previously carried out, we now make assessment of the high-frequency loss at the tenth harmonic, or 1.5 MHz, as follows:

$$\alpha = 20\log_{10}\left[1+\frac{1500}{24}\right]^n \cong 72.0\,\text{dB}\Bigg|_{n=2} \tag{19.49}$$

And so, we have a common-mode filter that has approximately 72 dB of loss at 1.5 MHz. To ensure a defensive design, we may add a couple of additional Y capacitors to the design, and these are placed in parallel with the original. These are not to be populated and are there as a precaution should we need to increase the loss during testing. If the design uses a PCB chassis ground plane, these Y capacitors may be placed in areas on the PCB where current-steering techniques are employed to control current flow. These Y capacitors will help to shunt HF noise locally and prevent currents from taking unwanted paths within the copper planes.

19.4.6.9 Common-Mode Damping—dQ

The common-mode Y capacitors, along with the common-mode inductance, form a second-order filter and, once again, we must guard against the effects of resonance. If we assume tight coupling of the components and reasonable coupling of the choke, the dB gain at or around the cutoff frequency may be relatively high. To reduce the gain, RC dQ shunts may also be added in parallel with the common-mode Y capacitors in certain applications (Figure 19.16). This is a defensive design feature to ensure that the gain of the filter at the resonant frequency can be reduced to acceptable levels. The amplitude gain is due to interaction between the Y capacitors and the common-mode chokes, including parasitic inductance, in each line. If the design can stand to have these components added, they can be placed. However, they should not be populated unless testing shows that they are needed; better to be safe than sorry.

To size the common-mode RC shunt, use a resistor that is equal to the filter characteristic impedance, or $|Z_o|$. In this case, the impedance is

$$|Z_o| = \sqrt{(L_{CM} + L_p)/C_Y} \tag{19.50}$$

The term L_p is the dominant parasitic inductance. To select the shunt capacitor, simply define the resonant frequency f_r, of the problem frequency; then define C as follows:

$$C = \frac{1}{2\pi f_r |Z_o|} \tag{19.51}$$

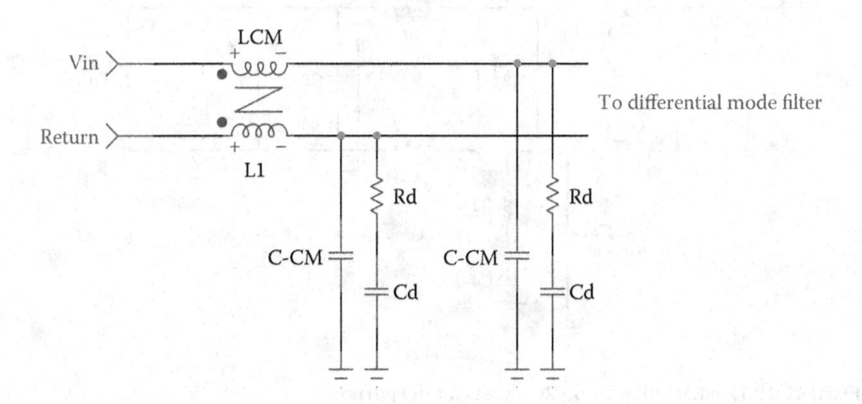

FIGURE 19.16 Common-mode filter with Y capacitor RC shunts.

19.4.6.10 Filter Design Summary

The filter structure in Figure 19.17 provides both common- and differential-mode loss. We have removed the series LR damping networks and included RC shunt dQ networks for the purposes of clarity. We should also mention that the common-mode RC shunts of Figure 19.17 are employed to reduce the circuit Q, and with a nominal inductance of 1.8 mH and a Y capacitor of 0.022 μF, the RC shunt terms are approximately 270 ohms and 0.1 μF, respectively. These are populated as needed during test.

The total differential mode capacitance is shown as C2. Both C1 and C3 are shown as optional and will, ofcourse, ensure adjustment in the -3dB pole Q frequency, therefore, allowing the value of L to be smaller if required. The design example calls out for 22-μF differential-mode capacitance, which may be impractical for many applications.

This filter is typically employed on DC power inputs and may also be used on single-phase AC power inputs as long as the X and Y capacitors are voltage rated correctly. With an AC filter, it is important to maintain adequate separation between the AC line frequency and the filter corner frequency; therefore, for AC applications, we would move the pole-Q frequency out to the right as far as is practicable while also maintaining sufficient loss. This may also drive the need for a four-pole section.

Figure 19.18 is the differential-mode response using the 3-ohm load and 50-ohm LISN. From a noise perspective, the 3-ohm impedance represents the power supply impedance, which is actually the source, and the 50-ohm LISN becomes the load for measurement purposes. Throughout this chapter, we have discussed specific corner frequencies, frequency-magnitude slopes of −40 dB per decade, etc. In reality, the source

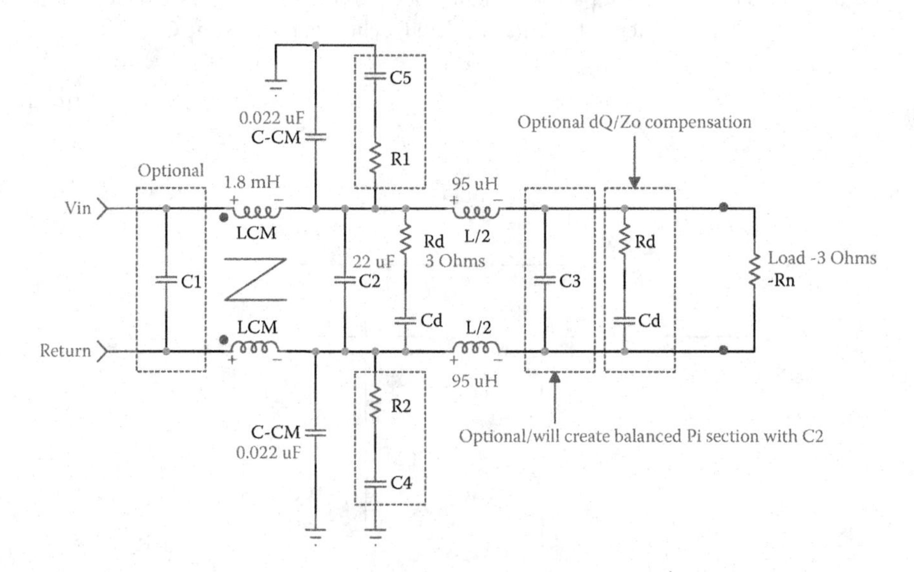

FIGURE 19.17 EMI filter using RC shunts for dQ purposes.

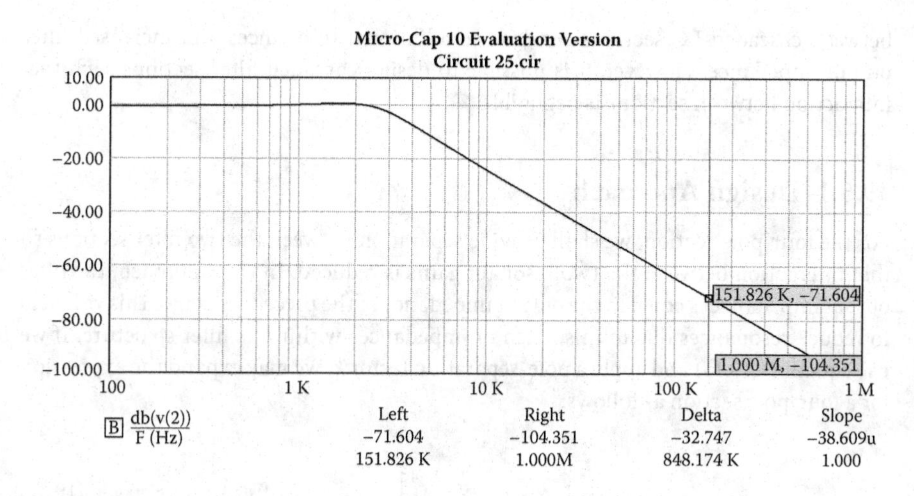

FIGURE 19.18 Frequency-magnitude slope—differential mode.

impedance will not be 50 ohms, and the load impedance may very well be 3 ohms, or even 33 ohms. So what we are saying is this: The performance of the filter is very much dictated by the insertion loss at the fundamental harmonic, and this is driven by the slope of the response (dB/decade) and the −3-dB pole-Q frequency. More loss equates to more poles. Furthermore, control of filter Q (damping) and, of course, the source and load impedances will also dictate the final response. Once again, this is an EMI filter and not a high-performance linear-phase filter with a perfectly defined −3-dB pole-Q frequency. As long as the filter has the insertion loss needed to bring the fundamental harmonic below the dB amplitude limit, then passing EMC testing should not prove to be a show stopper. Some tuning of the filter may be necessary, and if the design has the defensive attributes we stated, then filter adjustment should bring a successful result to testing.

19.5 Four-Pole LC Structure

This section of the chapter will extend the two-pole LC filter design process to a four-pole LC structure to show component optimization. Power converters are becoming more compact and are running at higher frequencies. The need for smaller filter components is essential, both for packaging and form factor needs as well as meeting the requirements for weight, etc. The four-pole filter is designed in a similar fashion to that of the two-pole structure with the exception that each two-pole structure has its own resonant frequency. Furthermore, the cascade connection of two LC filter sections can achieve a given high-frequency attenuation with less volume and weight than a single-section LC filter. The impact to this is that the pole-Q frequency of the multiple-section filter allows use of smaller inductance and capacitance values. Damping of each LC section is usually required, which implies that damping of each section should be optimized. Interactions between cascaded LC sections can lead to additional resonances and increased filter output impedance. However, it is possible to design cascaded filter sections such that interaction between sections is negligible.

19.5.1 Design Approach

For the four-pole section, we shall provide separation between the two filter sections so that interaction between the two resonant gains is reduced. In this approach, the filter output impedance is equal to the output impedance of the preceding stage. This will help to reduce resonances due to mismatch of impedances within the filter structure. If we use equation (19.18) and apply a pole-separation factor k, we may expand the expression for a four-pole section as follows:

$$\alpha = 20\log_{10}\left[\left(\frac{\omega_S}{\omega_C}\right)^2\left(\frac{\omega_S}{k\omega_C}\right)^2\right] = 20\log_{10}\left[\frac{\omega_S^4}{k^2\omega_C^4}\right] \tag{19.52}$$

We want to keep the pole-Q frequencies for each LC section apart to reduce interaction between stages; therefore, for the design example, we shall use $k = 3$. We also know that the attenuation α needed is –72 dB. Therefore, the attenuation factor A is $1/[10^{-(72/20)}] = 3981$. To find the first frequency, we solve for ω_C in equation (19.53).

$$\omega_C = \left[\frac{\omega_S{}^2}{k \times 63}\right]^{0.5} = f_1 \cong 10.9kHz \tag{19.53}$$

This is the base frequency and is used for the first LC structure. To realize the second upper frequency, we simply multiply f_1 by k to get $f_2 = 32.7$ kHz.

The impedance of each filter section must be equal to, or lower than, the negative impedance of the converter of 3 ohms to maintain stability. To ensure stability for the four-pole section, we shall maintain each LC section at 3 ohms. If we use the impedance at 3 ohms along with the frequencies for $f_1 = 10.9$ kHz and $f_2 = 32.7$ kHz, we are able to define the values for L and C in each case.

$$L_1 = \frac{Z_O}{2\pi f_1} = 44\mu H \qquad L_2 = \frac{Z_O}{2\pi f_2} = 15\mu H$$

$$C_1 = \frac{1}{2\pi f_1 Z_O} = 4.8\mu F \qquad C_2 = \frac{1}{2\pi f_2 Z_O} = 1.6\mu F$$

If we use a balanced L in each section, we would halve the inductor values, making them as follows: section (1) $L/2 = 22$ μH and section (2) $L/2 = 7.5$ μH. The capacitors would stay the same in each case.

Figures 19.19 and 19.20 show the equivalent-circuit and frequency-magnitude response for the four-pole filter. Damping of the filter is achieved by using an RC shunt.

The filter may also be dQ damped using the series LR technique, and this is usually a decision that is driven by the size of the damping capacitor C_d.

The final 4-pole filter structure is shown in Figure 19.21. Note that C1 in Figure 19.21 has been given a value of 4.7 μF which is a standard value. The four-pole filter offers a better component selection, with capacitor values that are much more practicable due to the higher pole-Q frequencies of 10.9 kHz and 32.7 kHz. The frequency separation method allows the two sections to be designed with, as a minimum, an octave between

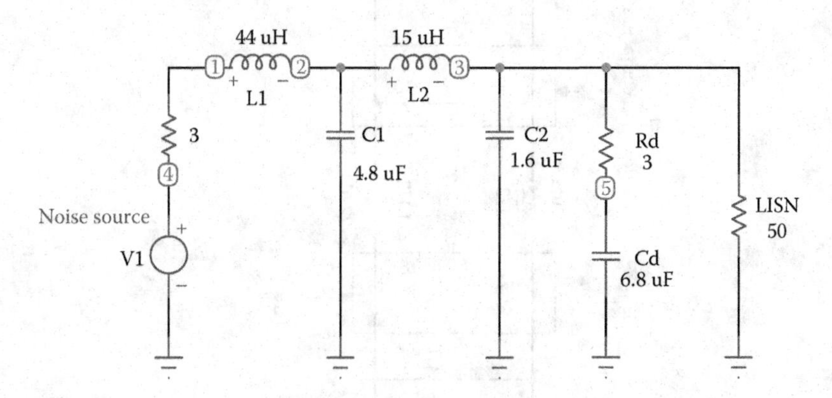

FIGURE 19.19 Four-pole filter structure with 50-ohm LISN and 3-ohm source.

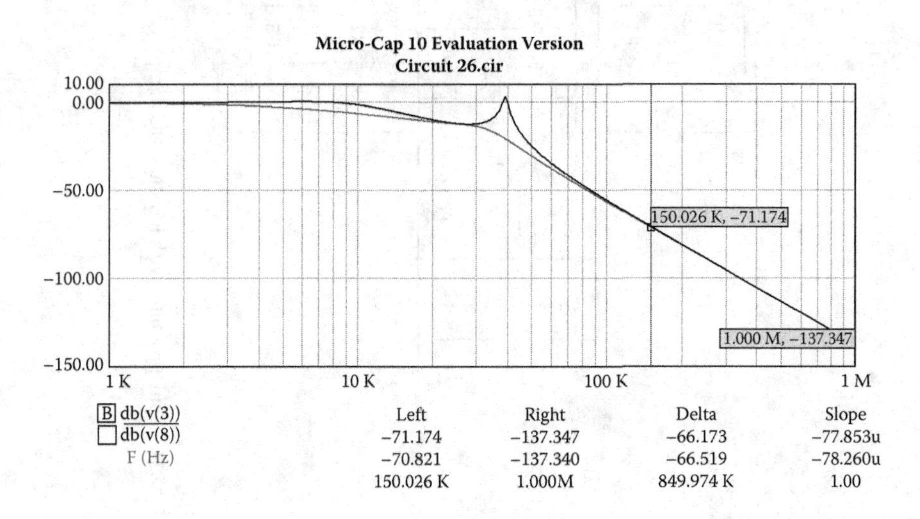

FIGURE 19.20 Frequency-magnitude slope of four-pole filter (with and without dQ damping).

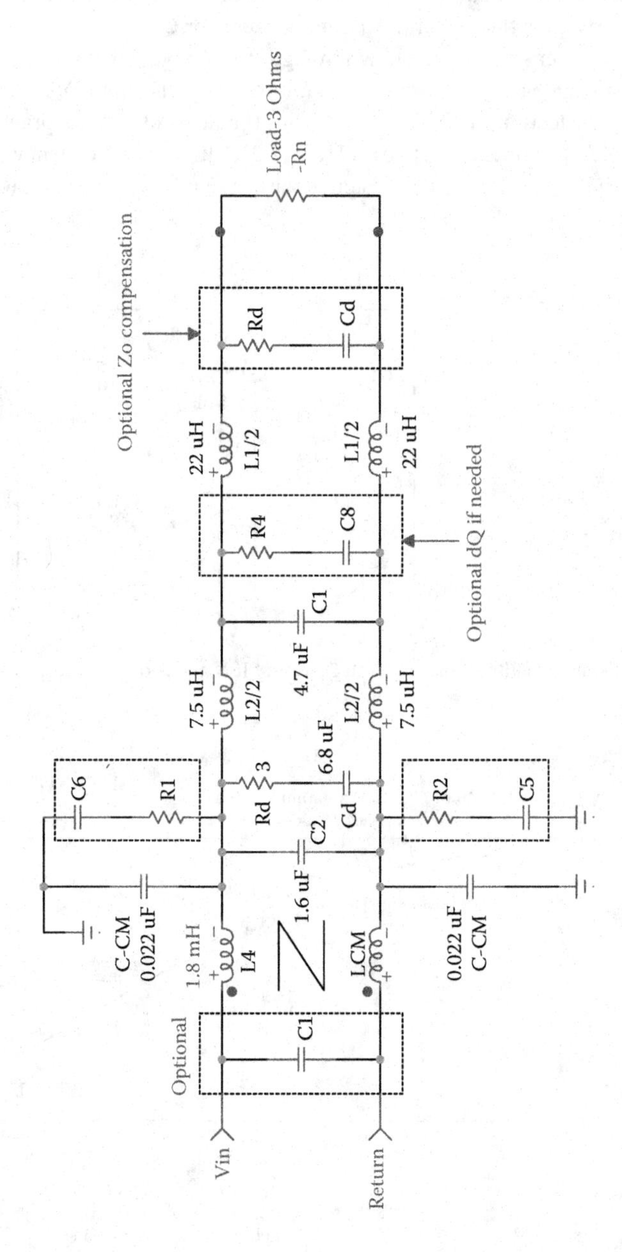

FIGURE 19.21 Four-pole filter with RC shunt dQ damping.

f_1 and f_2 so that the resonant frequencies do not interact. The design methods, both in this chapter and in chapter 16 using the *K* factors, offer a solution for EMI filter design, but again, there is no perfect solution to any filter. The main considerations are insertion loss versus frequency and making sure that the filter is able to handle the load current without core saturation. Core saturation is catastrophic for an EMI filter, as the inductance will drop considerably, thereby eliminating the insertion loss performance.

20

Packaging Information

Packaging of EMI filters is a very important subject and is often critical to the performance of the filter and the rest of the electronics that it is protecting. Component placement, proximity, and layout is important in all fields of electronics. However, components that are susceptible to *H* fields must be moved away from these fields, and wiring or printed circuit board traces must be as short as possible. There are two basic types of packaging solutions for EMI filters. These are: (a) directly connected components that are then potted and (b) the use of a PCB using conventional component placement techniques in order to reduce *H* field effects whilst maintaining low-inductance interconnects. These criteria present many challenges for packaging of EMI solutions, and even more so when today's designs are driving smaller form factors and reduced weight.

20.1 Layout

The physical layout of the EMI filter for best performance is an enclosure that is long and thin. The full length is much greater than the height and width, as shown in Figure 20.1.

In the EMI filter placement shown in Figure 20.2, the toroid components are spaced apart by the capacitors. This removes the tendency toward cross talk or mutual inductance by increasing the distance between the inductive components. Another method is by quadrature, as seen in the same figure.

Also, note in Figure 20.1 that the distance between both the input (dirty) and output (clean) extend the length of the enclosure where the output terminals are on the far end, away from the input connector end. In Figure 20.1, the output terminals are not shown, as they are on the blind side. If the input and output must be on the same face, a shield should run almost the full length of the filter, as shown in Figure 20.3. The components are divided between the two halves. Half run from the front to the back half and then double back to the front on the other side of the shield. The shield must be fully grounded for the entire length. This adds to the "fit" problem and may increase the width of the enclosure. These parameters still hold true for both low- and high-current filters. The filter's aspect ratio should be long in length compared with the height and width. The components should run from one end to the other, as in a transmission line, rather than hop back and forth within the filter body.

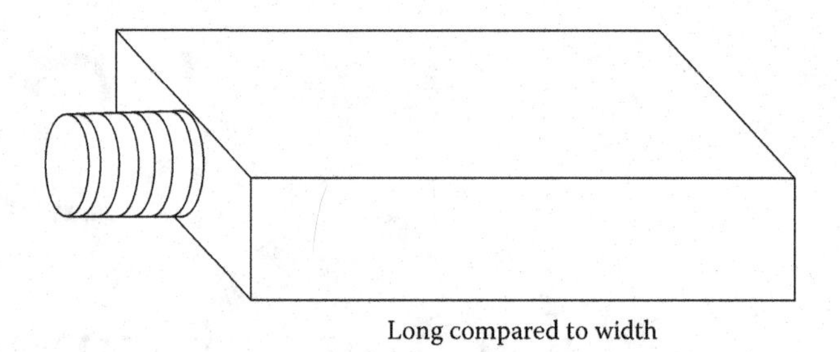

FIGURE 20.1 Width and height compared to length.

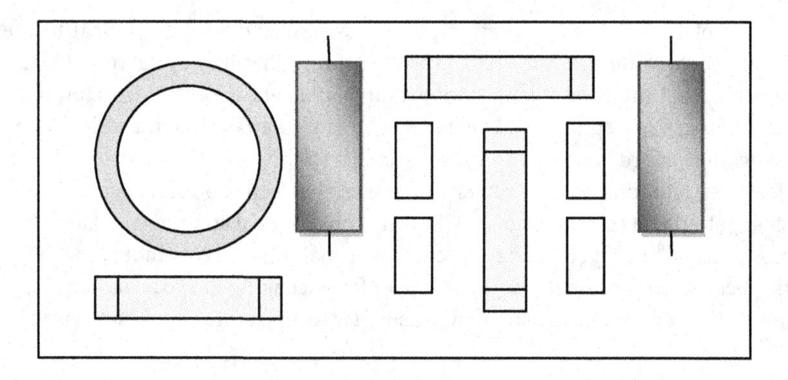

FIGURE 20.2 Toroids spaced apart and in quadrature.

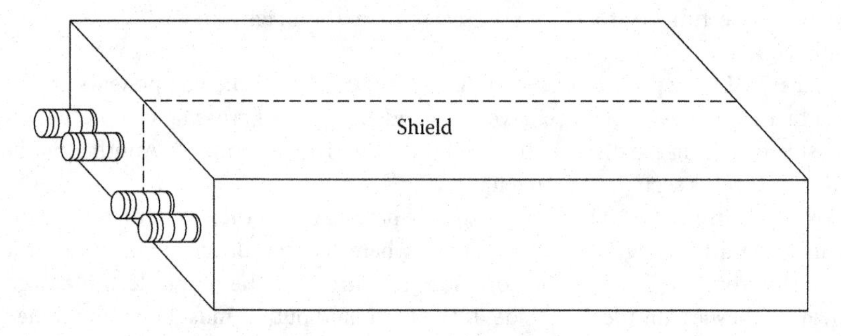

FIGURE 20.3 Shield separating input and output.

This reduces the cross-talk effect, and Capcon should be used to cover lead wires to help attenuate the upper frequencies. The inductors should be mounted in quadrature—including toroids—as shown in Figure 20.2. The alternative requires more room. Figure 20.2 shows two toroids in quadrature and two toroids that are not. However, they are separated by additional distance. The upper right and the lower left are in the same plane, with a capacitor between, but the distance is farther, which reduces the magnetic coupling. The top two inductors would be wired directly to the line-to-line capacitor and the bottom two also wired directly to the capacitor using the "vee" technique (long inductor leads wired directly to the capacitor). This continues to the next section, still maintaining the quadrature of the inductors. These are not usually mounted to a printed circuit board because of the current and the DC resistance of the boards.

Most EMI houses do not design their filters to utilize printed circuit boards because of trace parasitic effects (inductance) and the cumulative stray capacitance between power and ground planes. In certain applications, the PCB solution must be employed, but careful control of copper, trace inductance, and placement is critical to ensure EMC performance. In certain cases, an EMI filter may have been designed as part of the power input, and this is acceptable if the filter is mounted in a container that is a good conductor. Furthermore, the filter must be grounded directly to the ground plane so that the feed-through capacitors can function properly. This works well if the required insertion loss is for Federal Communications Commission (FCC) standards or if the required loss is low. Other design approaches place the filter within the power supply using open or exposed components. This technique rarely works without redesign or iterative adjustment. The filter must have shielded components to function properly. If this is not in place, adjacent magnetic fields either are influenced by the filter or couple their magnetic field to the inductors of the filter. As a consequence, a 60-dB filter, for example, is now only 24 dB, and the filter designer quite often has no idea why the filter has failed testing.

The case or container of the filter must be a good conductor. The better this surface conducts, the lower the magnetic field is on the outside of the case. Even though cold-rolled steel is often used for the filter body, the container is often silver plated to enhance the conduction on the inside and outside of the enclosure. This is best for military applications or other groups requiring severe loss. This approach attenuates the H field and improves the radiated emissions.

The H field establishes a current on the surface of the filter body container due to the low resistance of the plating. The better this surface conducts and the thicker the case material, the weaker is the current on the other side of the case. This reduces the H field departing from the case wall and is true for H fields propagating either in or out of the filter. The container must also be a good conductor so that the feed-through capacitors can function. The same would be true if the "good" conducting case was not wired to ground or if the ground wire was loose or missing. In the circuit of Figure 20.4, neither the four feed-through capacitors nor the two grounded common-mode arresters work properly if the case ground is resistive, if the container has a high resistance, or if the ground lead is not low inductance. The one thing that we can say about ground is that it isn't ground, or as most say, *ground isn't*. So, grounding of these filters cannot be marginal or overlooked.

20.2 Estimated Volume

Col. W. T. McLyman has provided the data in Table 20.1 to calculate many different magnetic properties from his *Transformer and Inductor Design Handbook* (Marcel Dekker, New York). This book is a "must have" for electrical engineers and anyone involved in EMI filter design. Our goal here is to determine the volume needs of the filter and, more

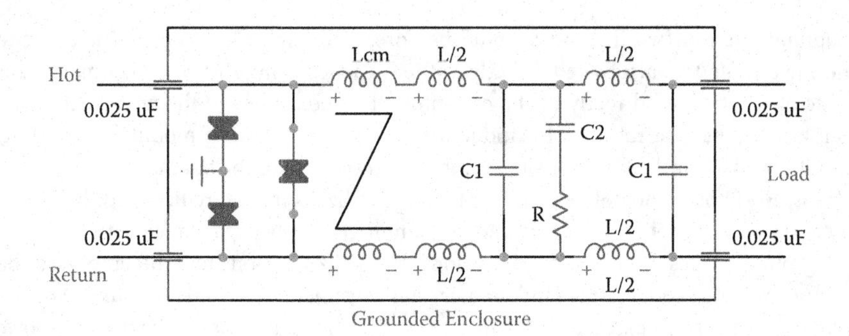

FIGURE 20.4 Grounding—MOVs and four feed-throughs may be ungrounded.

TABLE 20.1 McLyman Magnetic Fractions

Core Type	K_j (at 25°C)	K_j (at 50°C)	X	K_v (cm³)	K_w (g)
Pot core	433	632	1.20	14.5	48.0
Power core	403	590	1.14	13.1	58.8
Laminations	366	534	1.14	19.7	68.2
C core	323	468	1.16	17.9	66.6
Single coil	395	569	1.16	25.6	76.6
Tape wound	250	365	1.15	25.0	82.3

Source: Colonel W. T. McLyman. 2004. *Transformer and Inductor Design Handbook 3rd ed.*, New York: Marcel Dekker. With permission.

importantly, to determine an accurate assessment of weight (discussed later in this section). The main idea is to find the area product (A_p) in centimeters to the fourth power. From this, and knowing the core, the approximate volume can be found based upon the filter structure. According to McLyman in the method listed in his book, the energy of the inductor must be determined. The equation is

$$E = \frac{LI_p^2}{2} \tag{20.1}$$

The type of core to be used must be known in order to use Table 20.1. The values for K_j and X also come from Table 20.1. K_u is the winding factor—0.4 for a toroid—and B_m (in teslas) must be known for the core type. The A_p is

$$A_p = \frac{|2(E) \times 10^4|^x}{B_m K_u K_j} \tag{20.2}$$

Find the A_p for the different sizes of inductors and add the different A_p values for all the inductors for the total A_p. For example, a three-stage balanced L filter would require six inductors, all of the same value. Find the energy based on the peak current, and from knowledge of the core type, obtain the components for the A_p.

If the peak current is 5 A and the inductors are 250 μH, then E is

$$E = \frac{250 \times 10^{-6} \times 5^2}{2} = 0.0031 \tag{20.3}$$

If MPP powder cores are the choice, then $K_u = 0.4$, K_j at 25°C = 403, $B_m = 0.7$ tesla, and $X = 1.14$. The A_p follows:

$$A_p = \frac{|2 \times 0.0031 \times 10^4|^{1.14}}{0.7 \times 0.4 \times 403} = 0.9792 \tag{20.4}$$

Round this up to 1 and find the total A_p and then the volume

$$VL_{tot} = 1 \times 13.1 \times 6 = 78.6cm^3 \tag{20.5}$$

In equation (20.5), the value 13.1 comes from Table 20.1 for the MPP core in cubic centimeters. The capacitors are not included, but the weight ratio of the capacitors is on the order of that of the inductors. The total would be 160 cm³, and this volume is only 60% utilized. This gives 267 cm³, but allow for the container, feed-throughs, and wiring; round this up to 280 cm³ (17 in.³). These are rough estimates that should get the design engineer into the ballpark for the inductors. For the capacitors that are round, the outside diameter squared times the height works. Then multiply by the number. For the pressed-type capacitors, or those that are not round, we shall use length × height × width. This applies to all other components. This is also true for Colonel McLyman's inductors.

20.3 Volume-to-Weight Ratio

Bob Hassett, chief engineer at RFI Corp. (now retired), has carried out research on the size-to-weight ratio for EMI filters. This was primarily done on the tubular types of filters mentioned in an earlier chapter. The ratio is 1.5 ounces per cubic inch. This equates to 1.6 pounds for the filter discussed in the preceding section, which needed 17 in.³ (280 cm³). From McLyman, again use his area product and use the same A_p as in the preceding section for one inductor. Multiplying this value of 58.8 for each inductor by the six inductors, as listed for the powder cores in grams, gives

$$58.8 \times 6 = 352.8 \text{ g}$$

Doubling this for the capacitors, wiring, feed-through capacitors, container, and input terminals, 352.8 × 2 = 705.6 g. Thus, the weight in pounds is approximately

$$\frac{705.6}{454} = 1.55 \text{ lb}$$

At Hopkins Engineering located in Sylmar, California (now closed), research was carried out on both volume and weight in the late 1980s. This follows on from the information above as in the case of a feed-through capacitor; thus we may use $O_d^2(length)$. This holds true for components that are flanged with solder pins. Length here means from pin end to pin end. Round units, such as capacitors, use the same formula, and the height would include any terminals. Square units like transformers are simply L × W × H. Small units like a MOV or a low-wattage resistor are just thrown in. Take the total and divide by 0.6.

20.4 Potting Compounds

Potting compounds add substantial weight and so should be used sparingly. It is often considered that potting of EMI filters enhance cooling. However, most potting compounds do not aid this function. In fact, they often hinder heat transfer. This could be a desirable feature by avoiding heat transfer to sensitive components. If heat transfer is the goal, some compounds have this feature, but they are often too expensive or not readily available. Often, fine granules of aluminum are added to the epoxy to enhance the heat transfer. Most often, potting is added to the filter to tie down the components. It is important not to fill the enclosure entirely with the potting material but only cover the bottom to a height necessary to support the components.

Appendix A: *K* Values of Different Topologies

These tables provide a simple method of deriving the cut-off frequency for a defined filter structure and number of poles. Use of the tables is as follows.

Using Table A.2 select the topology such as a double π. Look down the dB column to find the loss in dB required. For example, a loss of 40 dB is required. The value close to 40 dB is listed as 40.98. Follow over to the left hand column where the *K* value = 4. Assume that the frequency requiring this loss is, for the purposes of example, 160 kHz and divide this by 4.

In this example, 160 kHz/4 = 40 kHz. Therefore, 40 kHz is the −3-dB cutoff, or pole-Q frequency, and the values of *L* and *C* can be calculated as follows:

$$\mathrm{L} = \frac{R_\mathrm{d}}{2\pi 40{,}000} \qquad \mathrm{C} = \frac{\mathrm{L}}{R_\mathrm{d}^2}$$

where R_d is either the design impedance, or 50 ohms, or whatever the specification demands.

Single Filter Structures

Single L filter

$$\mathrm{dB} = 20\log_{10}\left[\frac{(4+K^4)^{0.5}}{2}\right] \tag{A.1}$$

Single π filter

$$dB = 20\log_{10}\left[\frac{1}{2}\sqrt{\frac{64+K^6}{16}}\right] \quad (A.2)$$

Single T filter

$$dB = 20\log_{10}\left[\frac{1}{2}\sqrt{\frac{64+K^6}{16}}\right] \quad (A.3)$$

Double Filter Structures

Double L filter

$$dB = 20\log_{10}\left[\frac{1}{2}\sqrt{4+4K^4-4K^6+K^8}\right] \quad (A.4)$$

Double π filter

$$dB = 20\log_{10}\left[\frac{1}{2}\sqrt{\frac{64+4K^6-4K^8+K^{10}}{16}}\right] \quad (A.5)$$

Double T filter

$$dB = 20\log_{10}\left[\frac{1}{2}\sqrt{\frac{64+4K^6-4K^8+K^{10}}{16}}\right] \quad (A.6)$$

Triple Filter Structures

Triple L filter

$$dB = 20\log_{10}\left[\frac{1}{2}\sqrt{4+9K^4 124K^8-8K^{10}+K^{12}}\right] \quad (A.7)$$

Triple π filter

$$dB = 20\log_{10}\left[\frac{1}{2}\sqrt{\frac{64+9K^6-24K^8+22K^{10}-8K^{12}+K^{14}}{16}}\right] \quad (A.8)$$

Triple T filter

$$\text{dB} = 20\log_{10}\left[\frac{1}{2}\sqrt{\frac{64+9K^6-24K^8+22K^{10}-8K^{12}+K^{14}}{16}}\right] \quad (A.9)$$

Quad Filter Structures

Quad L filter

$$\text{dB} = 20\log_{10}\left[\frac{1}{2}\sqrt{4+16K^4-80K^6+148K^8-128K^{10}56K^{12}-12K^{14}+K^{16}}\right] \quad (A.10)$$

Quad π filter

$$\text{dB} = 20\log_{10}\left[\frac{1}{2}\sqrt{\frac{64+16K^6-80K^8+148K^{10}-128K^{12}+56^{14}-12K^{16}}{16}}\right]z \quad (A.11)$$

Quad T filter

$$\text{dB} = 20\log_{10}\left[\frac{1}{2}\sqrt{\frac{64+16K^6-80K^8+148K^{10}-128K^{12}+56^{14}-12K^{16}}{16}}\right] \quad (A.12)$$

Special Combinations

T and L filter

$$\text{dB} = 20\log_{10}\left[\frac{1}{2}\sqrt{\frac{64+16K^4-15K^6+9K^8}{16}}\right] \quad (A.13)$$

π + L filter

$$\text{dB} = 20\log_{10}\left[\frac{1}{2}\sqrt{\frac{64+16K^4+17K^6-9K^8+K^{10}}{16}}\right] \quad (A.14)$$

Single π capacitors at twice value

$$\text{dB} = 20\log_{10}\left[\frac{1}{2}\sqrt{4+K^2-2K^4+K^6}\right] \quad (A.15)$$

TABLE A.1 Single Structure K values

K	Single L	20 log dB	Single π	20 log dB	Single T	20 log dB
3.0	4.61	13.27	3.52	10.93	3.52	10.93
3.1	4.91	13.82	3.86	11.72	3.86	11.72
3.2	5.22	14.35	4.22	12.50	4.22	12.50
3.3	5.54	14.86	4.60	13.26	4.60	13.26
3.4	5.87	15.37	5.01	14.00	5.01	14.00
3.5	6.21	15.86	5.45	14.73	5.45	14.73
3.6	6.56	16.33	5.92	15.44	5.92	15.44
3.7	6.92	16.80	6.41	16.14	6.41	16.14
3.8	7.29	17.25	6.93	16.82	6.93	16.82
3.9	7.67	17.70	7.48	17.48	7.48	17.48
4.0	8.06	18.13	8.06	18.13	8.06	18.13
4.1	8.46	18.55	8.67	18.76	8.67	18.76
4.2	8.88	18.96	9.31	19.38	9.31	19.38
4.3	9.30	19.37	9.99	19.99	9.99	19.99
4.4	9.73	19.76	10.69	20.58	10.69	20.58
4.5	10.17	20.15	11.43	21.16	11.43	21.16
4.6	10.63	20.53	12.21	21.73	12.21	21.73
4.7	11.09	20.90	13.02	22.29	13.02	22.29
4.8	11.56	21.26	13.86	22.84	13.86	22.84
4.9	12.05	21.62	14.74	23.37	14.74	23.37
5.0	12.54	21.97	15.66	23.89	15.66	23.89
5.1	13.04	22.31	16.61	24.41	16.61	24.41
5.2	13.56	22.64	17.60	24.91	17.60	24.91
5.3	14.08	22.97	18.64	25.41	18.64	25.41
5.4	14.61	23.30	19.71	25.89	19.71	25.89
5.5	15.16	23.61	20.82	26.37	20.82	26.37
5.6	15.71	23.92	21.97	26.84	21.97	26.84
5.7	16.28	24.23	23.17	27.30	23.17	27.30
5.8	16.85	24.53	24.41	27.75	24.41	27.75
5.9	17.43	24.83	25.69	28.20	25.69	28.20
6.0	18.03	25.12	27.02	28.63	27.02	28.63

TABLE A.2 Double Structure K values

K	Double L	20 log dB	Double π	20 log dB	Double T	20 log dB
3.0	31.52	29.97	23.65	27.48	23.65	27.48
3.1	36.58	31.26	28.36	29.05	28.36	29.05
3.2	42.20	32.51	33.77	30.57	33.77	30.57
3.3	48.42	33.70	39.95	32.03	39.95	32.03
3.4	55.27	34.85	46.98	33.44	46.98	33.44
3.5	62.79	35.96	54.94	34.80	54.94	34.80
3.6	71.03	37.03	63.93	36.11	63.93	36.11
3.7	80.02	38.06	74.02	37.39	74.02	37.39
3.8	89.82	39.07	85.33	38.62	85.33	38.62
3.9	100.47	40.04	97.96	39.82	97.96	39.82
4.0	112.00	40.98	112.00	40.98	112.00	40.98
4.1	124.48	41.90	127.59	42.12	127.59	42.12
4.2	137.95	42.79	144.85	43.22	144.85	43.22
4.3	152.45	43.66	163.89	44.29	163.89	44.29
4.4	168.05	44.51	184.85	45.34	184.85	45.34
4.5	184.78	45.33	207.88	46.36	207.88	46.36
4.6	202.72	46.14	233.12	47.35	233.12	47.35
4.7	221.90	46.92	260.73	48.32	260.73	48.32
4.8	242.38	47.69	290.86	49.27	290.86	49.27
4.9	264.23	48.44	323.68	50.20	323.68	50.20
5.0	287.50	49.17	359.38	51.11	359.38	51.11
5.1	312.25	49.89	398.12	52.00	398.12	52.00
5.2	338.54	50.59	440.10	52.87	440.10	52.87
5.3	366.44	51.28	485.53	53.72	485.53	53.72
5.4	395.99	51.95	534.59	54.56	534.59	54.56
5.5	427.28	52.61	587.51	55.38	587.51	55.38
5.6	460.37	53.26	644.51	56.18	644.51	56.18
5.7	495.31	53.90	705.82	56.97	705.82	56.97
5.8	532.19	54.52	771.67	57.75	771.67	57.75
5.9	571.06	55.13	842.31	58.51	842.31	58.51
6.0	612.00	55.74	918.00	59.26	918.00	59.26

TABLE A.3 Triple Structure K values

K	Triple L	20 log dB	Triple π	20 log dB	Triple T	20 log dB
4.0	1560.00	63.86	1560.00	63.86	1560.00	63.86
4.1	1835.12	65.27	1880.99	65.49	1880.99	65.49
4.2	2148.64	66.64	2256.07	67.07	2256.07	67.07
4.3	2504.66	67.97	2692.51	68.60	2692.51	68.60
4.4	2907.58	69.27	3198.34	70.10	3198.34	70.10
4.5	3362.13	70.53	3782.40	71.56	3782.40	71.56
4.6	3873.40	71.76	4454.41	72.98	4454.41	72.98
4.7	4446.81	72.96	5225.00	74.36	5225.00	74.36
4.8	5088.17	74.13	6105.81	75.71	6105.81	75.71
4.9	5803.70	75.27	7109.53	77.04	7109.53	77.04
5.0	6600.00	76.39	8250.00	78.33	8250.00	78.33
5.1	7484.12	77.48	9542.25	79.59	9542.25	79.59
5.2	8463.54	78.55	11002.60	80.83	11002.60	80.83
5.3	9546.22	79.60	12648.74	82.04	12648.74	82.04
5.4	10740.58	80.62	14499.79	83.23	14499.79	83.23
5.5	12055.57	81.62	16576.41	84.39	16576.41	84.39
5.6	13500.63	82.61	18900.88	85.53	18900.88	85.53
5.7	15085.76	83.57	21497.21	86.65	21497.21	86.65
5.8	16821.51	84.52	24391.19	87.74	24391.19	87.74
5.9	18719.01	85.45	27610.54	88.82	27610.54	88.82
6.0	20790.00	86.36	31185.00	89.88	31185.00	89.88
6.1	23046	87.25	35146	90.92	35146	90.92
6.2	25503	88.13	39528	91.94	39528	91.94
6.3	28171	89.00	44386	92.94	44386	92.94
6.4	31066	89.85	49705	93.93	49705	93.93
6.5	34203	90.68	55579	94.90	55579	94.90
6.6	37597	91.50	62035	95.85	62035	95.85
6.7	41266	92.31	69121	96.79	69121	96.79
6.8	45227	93.11	76885	97.72	76885	97.72
6.9	49497	93.89	85382	98.63	85382	98.63
7.0	54066	94.66	94668	99.52	94668	99.52

TABLE A.4 Quad Structure K values

K	Quad L	20 log dB	Quad π	20 log dB	Quad T	20 log dB
.03	1481	63.41	2196	66.83	2196	66.83
3.1	2044	66.21	2900	69.25	2900	69.25
3.2	2780	68.88	3794	71.58	3794	71.58
3.3	3729	71.43	4919	73.84	4919	73.84
3.4	4940	73.87	6322	76.02	6322	76.02
3.5	6470	76.22	8063	78.13	8063	78.13
3.6	8389	78.47	10206	80.18	10206	80.18
3.7	10775	80.65	12828	82.16	12828	82.16
3.8	13720	82.75	16020	84.09	16020	84.09
3.9	17330	84.78	19881	85.97	19881	85.97
4.0	21728	86.74	24528	87.79	24528	87.79
4.1	27054	88.64	30095	89.57	30095	89.57
4.2	33467	90.49	36732	91.30	36732	91.30
4.3	41149	92.29	44610	92.99	44610	92.99
4.4	50308	94.03	53922	94.64	53922	94.64
4.5	61174	95.73	64885	96.24	64885	96.24
4.6	74012	97.39	77742	97.81	77742	97.81
4.7	89114	99.00	92766	99.35	92766	99.35
4.8	106813	100.57	110262	100.85	110262	100.85
4.9	127475	102.11	130569	102.32	130569	102.32
5.0	151513	103.61	154064	103.75	154064	103.75

TABLE A.5 Special Structure K values

K	Mix L	20 log dB	Mix π	20 log dB	Mix T	20 log dB
3.0	27.80	28.88	14.66	23.32	12.04	21.61
3.1	31.87	30.07	18.47	25.33	13.38	22.53
3.2	36.36	31.21	22.92	27.21	14.82	23.42
3.3	41.30	32.32	28.09	28.97	16.35	24.27
3.4	46.73	33.39	34.05	30.64	17.98	25.10
3.5	52.67	34.43	40.87	32.23	19.71	25.90
3.6	59.16	35.44	48.65	33.74	21.55	26.67
3.7	66.23	36.42	57.47	35.19	23.50	27.42
3.8	73.90	37.37	67.44	36.58	25.56	28.15
3.9	82.22	38.30	78.66	37.91	27.73	28.86
4.0	91.22	39.20	91.22	39.20	30.02	29.55
4.1	100.93	40.08	105.25	40.44	32.43	30.22
4.2	111.39	40.94	120.87	41.65	34.96	30.87
4.3	122.64	41.77	138.19	42.81	37.62	31.51
4.4	134.72	42.59	157.36	43.94	40.40	32.13
4.5	147.66	43.39	178.51	45.03	43.32	32.73
4.6	161.51	44.16	201.79	46.10	46.38	33.33
4.7	176.30	44.92	227.35	47.13	49.57	33.90
4.8	192.08	45.67	255.34	48.14	52.91	34.47
4.9	208.89	46.40	285.94	49.13	56.38	35.02
5.0	226.77	47.11	319.32	50.08	60.01	35.56
5.1	245.78	47.81	355.65	51.02	63.78	36.09
5.2	265.95	48.50	395.13	51.93	67.71	36.61
5.3	287.33	49.17	437.94	52.83	71.80	37.12
5.4	309.96	49.83	484.30	53.70	76.04	37.62
5.5	333.91	50.47	534.42	54.56	80.44	38.11
5.6	359.20	51.11	588.51	55.40	85.01	38.59
5.7	385.91	51.73	646.80	56.22	89.75	39.06
5.8	414.07	52.34	709.53	57.02	94.66	39.52
5.9	443.73	52.94	776.95	57.81	99.74	39.98
6.0	474.96	53.53	849.30	58.58	105.00	40.42

Appendix B: LC Passive Filter Design

TABLE B.1 Butterworth Normalized Coefficients

Order	C1	L2	C3	L4	C5	L6	C7	L8	C9	L10
1	2.000									
2	1.41421	1.41421								
3	1.00000	2.00000	1.00000							
4	0.76537	1.84776	1.84776	0.76537						
5	0.61803	1.61803	2.00000	1.61803	0.61803					
6	0.51764	1.41421	1.93185	1.93185	1.41421	0.51764				
7	0.44504	1.24698	1.80194	2.00000	1.80194	1.24698	0.44504			
8	0.39018	1.11114	1.66294	1.96157	1.96157	1.66294	1.11114	0.39018		
9	0.34730	1.00000	1.53209	1.87938	2.00000	1.87938	1.53209	1.00000	0.34730	
10	0.31287	0.90798	1.41421	1.78201	1.97538	1.97538	1.78201	1.41421	0.90798	0.31287
	L1	C2	L3	C4	L5	C6	L7	C8	L9	C10

Note: Table B.1 lists prototype element values for the normalized low-pass function, which assumes a cutoff frequency of 1 radian/second and source and load impedances of 1 Ω. Either an input capacitor (top reference line in table; see Figure B.1) or an input inductor (bottom line in table; see Figure B.2) can be used. These values are normalized scaling terms and may be used to design filters that require a Butterworth amplitude response.

TABLE B.2 Butterworth—Normalized Poles

Order (n)	Real σ	Imaginary $j\omega$
1	1.0000	
2	0.7071	0.7071
3	0.5000	0.8660
	1.0000	
4	0.9239	0.3827
	0.3827	0.9239
5	0.8090	0.5878
	0.3090	0.9511
	1.0000	
6	0.9659	0.2588
	0.7071	0.7071
	0.2588	0.9659
7	0.9010	0.4339
	0.6235	0.7818
	0.2225	0.9749
	1.0000	
8	0.9808	0.1951
	0.8315	0.5556
	0.5556	0.8315
	0.1951	0.9808
9	0.9397	0.3420
	0.7660	0.6428
	0.5000	0.8660
	0.1737	0.9848
	1.0000	
10	0.9877	0.1564
	0.8910	0.4540
	0.7071	0.7071
	0.4540	0.8910
	0.1564	0.9877

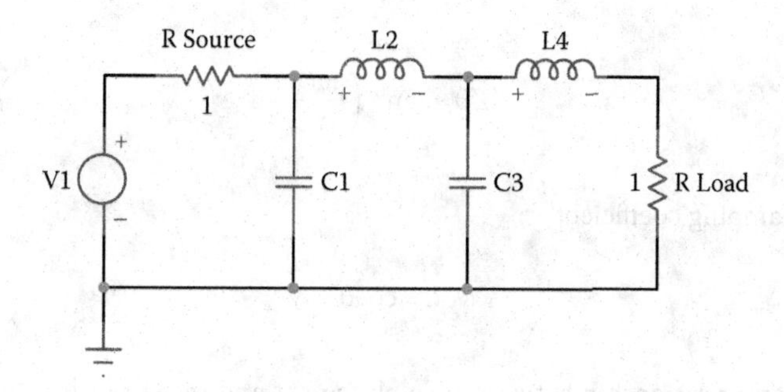

FIGURE B.1 Capacitor input.

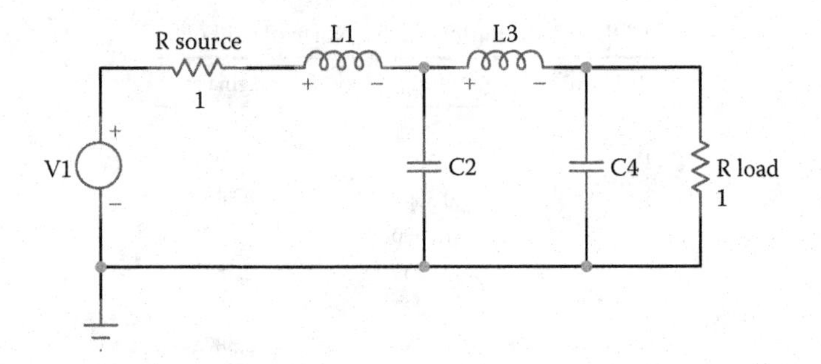

FIGURE B.2 Inductor input.

General Quadratic Form for Second-Order Homogeneous System

Transfer function—second-order system

$$CE = \frac{d^2y}{dt^2} + 2\zeta\omega_n \frac{dy}{dt} + \omega_n^2 y = 0$$

Poles of the characteristic equation

$$CE = \frac{1}{s^2 + 2\zeta\omega_n s + \omega_n^2} \triangleq \frac{1}{(s - p_1)(s - p_2)} \rightarrow p_1, p_2 = -\zeta\omega_n \pm j\omega_n\sqrt{1-\zeta^2}$$

Damped frequency of oscillation

$$\omega_d = \omega_n\sqrt{1-\zeta^2}$$

Define theta

$$\angle\theta = \tan^{-1}\left(\frac{j\omega}{\sigma}\right)$$

Define damping coefficient

$$\zeta = \cos\theta$$

S-plane and relationship between pole placement and second-order characteristics (Figure B.3)

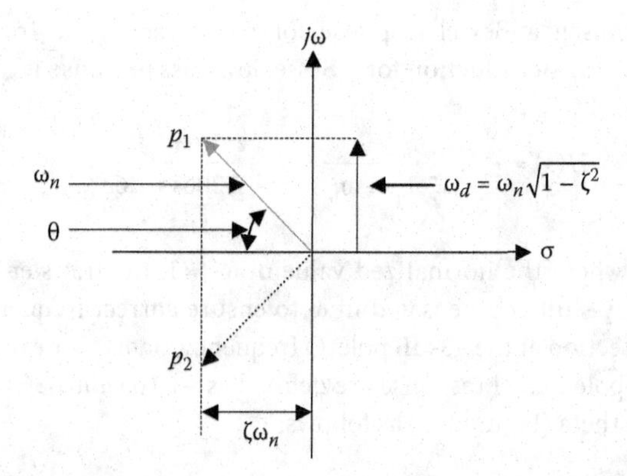

FIGURE B.3 S-plane.

General Quadratic Form for Second-Order Low-Pass Filter

$$H(s) = \frac{\omega_n^2}{s^2 + 2\zeta\omega_n s + \omega_n^2} \triangleq \frac{\omega_n^2}{s^2 + \frac{\omega_n}{Q}s + \omega_n^2} \Rightarrow \left|2\zeta\omega_n \equiv \frac{\omega_n}{Q}\right|$$

Damping factor

$$Q = \frac{1}{2\zeta}$$

Complex poles as a function of Q

$$\frac{1}{(s-p_1)(s-p_2)} \rightarrow p_1, p_2 = -\frac{\omega_n}{2Q} \pm j\omega_n\sqrt{1-\left(\frac{1}{2Q}\right)^2}$$

Normalized Butterworth transfer function (two poles)

$$H(s) = \left.\frac{1}{s^2 + 1.414s + 1}\right|_{\omega_n^2 = 1}$$

Normalized poles for the Butterworth response

$$p_1, p_2 = -0.707 \pm j0.707$$

Normalized Bessel transfer function (two poles)

For clarity, we wanted to show contrast between some other amplitude response and that of a Butterworth response, where $\omega_n = 1$ for a normalized solution.

We have chosen a Bessel response for comparison. The frequency-shifted normalized transfer function for a Bessel low pass response is

$$H(s) = \frac{\omega_n^2}{s^2 + 2\zeta\omega_n s + \omega_n^2} = \left.\frac{1.63}{s^2 + 2.206s + 1.63}\right|_{\omega_n^2 = 1.63}$$

For any filter where the normalized value of $\omega_n \neq 1$, the transfer function must be frequency-shift compensated in ω to ensure correct frequency-magnitude slope intersection at the −3-dB pole-Q frequency, or ω_C. For example, the poles of the two-pole Bessel response are defined as $-1.103 \pm j0.6368$. From this, we may define theta (Figure B.3) as follows:

$$\measuredangle\theta = \tan^{-1}\left(\frac{j\omega}{\sigma}\right) = 30°$$

Therefore, the damping factor is $\zeta = \cos\theta = 0.866$ (slightly overdamped). The natural undamped frequency is

$$\omega_n = \sqrt{1.103^2 + 0.6368^2} = 1.273$$

For example, if we were to design a Bessel response for a −3-dB corner frequency of, say, 4 kHz or 25.13×10^3 rads/sec. we would frequency shift this frequency by 1.273, which is a constant term for a two-pole Bessel response. Therefore, we would actually design the filter and select the components based upon a −3-dB frequency of 5.092 kHz. The filter will then have a −3-dB corner frequency of 4.0 kHz. Note that the frequency shift factor is based upon the complex conjugate pole-pairs; therefore, in the case of a four-pole filter, there would be two different frequency shift factors (FSF), each based upon $FSF = \sqrt{j\omega^2 + \sigma^2}$ for each pole-pair.

TABLE B.3 Bessel Frequency Normalized Component Values

Order	Filter Component / Normalized Value (capacitance in μF; inductance in μH)
1	$C_1 = 2.00$
2	$C_1 = 2.14$
	$L_2 = 0.57$
3	$C_1 = 2.20$
	$L_2 = 0.97$
	$C_3 = 0.33$
4	$C_1 = 2.20$
	$L_2 = 1.08$
	$C_3 = 0.67$
	$L_4 = 0.23$

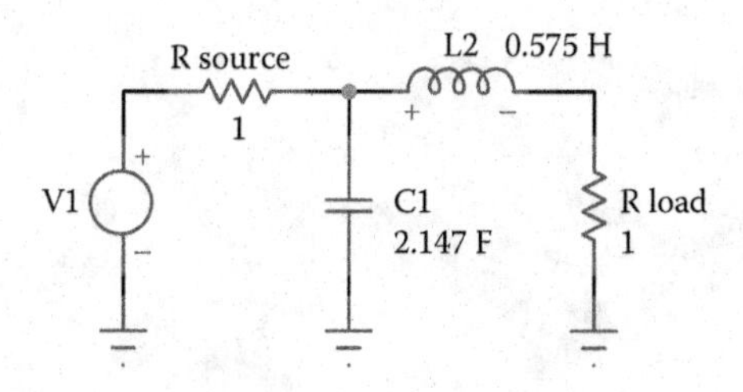

FIGURE B.4 Passive two-pole Bessel—normalized for ω = 1.

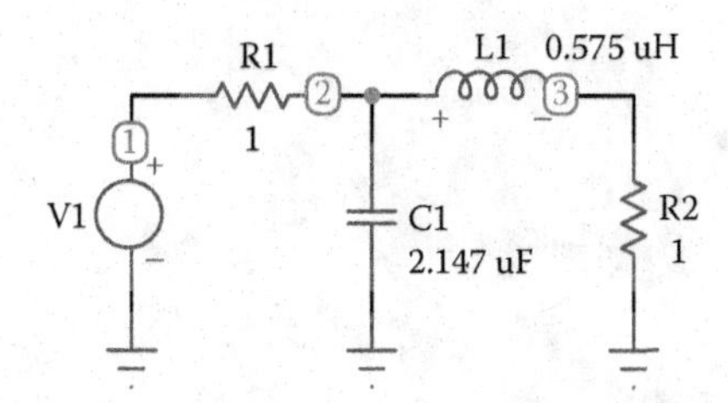

FIGURE B.5 Denormalized two-pole filter.

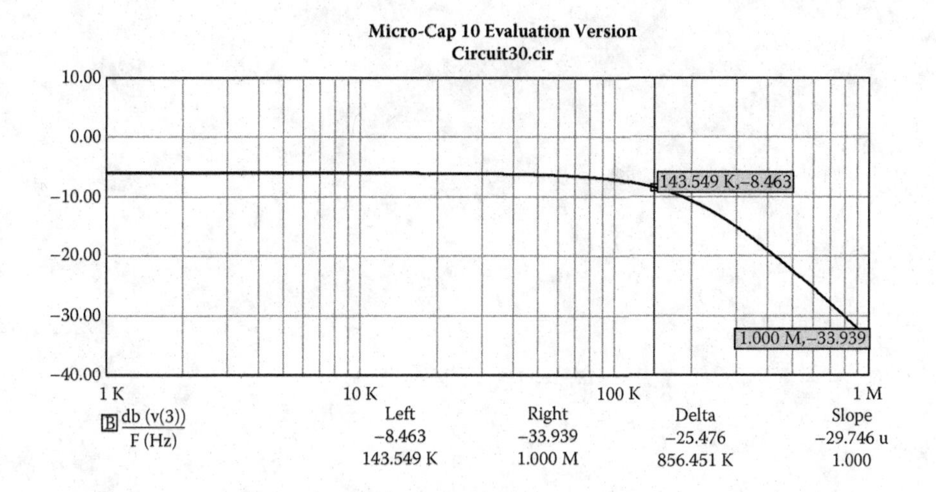

FIGURE B.6 Frequency magnitude loss for Figure B.5.

If we were to design a normalized Bessel filter with both source and load impedances of 1 ohm, we would use the frequency-normalized component scaling terms from Table B.3 [3].

If we denormalize the circuit of Figure B.4 and define the terms for *L* and *C* as 0.575 μH and 2.147 μF, respectively (Figure B.5), the frequency-magnitude response is shown in Figure B.6 with a flat-line loss of −6 dB and a corner frequency of 143.4 kHz.

Appendix C: Conversion Factors

Decibel Conversion for Power

Decibels relative to Power

$dB = 10\ \text{Log}_{10}(p_2/p_1)$

Decibel Conversion for Current

Decibels relative to Current

$dB = 20\ \text{Log}_{10}(A2/A1)$

Decibel Conversion for Microvolts (μV)

Decibels relative to 1 microvolt

$dB\mu V = 20\ \text{Log}_{10}(\text{voltage in V})/\mu V)$

Decibel Conversion for Microamps (μA)

Decibels relative to 1 microamp

$dB\mu A = 20\ \text{Log}_{10}(\text{current in A})/\mu A)$

V to dBμV

$$dB\mu V = 20\log_{10}(V) + 120$$

dBμV to V

$$V = 10^{((dB\mu V - 120)/20)}$$

dBV to dBμV

$$dB\mu V = dBV + 120$$

dBμV to dBV

$$dBV = dB\mu V - 120$$

dBμV to dBμA

$$dB\mu A = dB\mu V - 20\log_{10}(Z)$$

dBμA to dBμV

$$dB\mu V = dB\mu A + 20\log_{10}(Z)$$

Note: In the case of both $dB\mu V$ and $dB\mu A$, the term (Z) relates to impedance. Typically 50Ω.

References

1. Middlebrook, R. D. 1976. Input filter considerations in design and application of switching regulators. Paper presented at IEEE Industry Applications Society Annual Meeting, 366–382.
2. Middlebrook, R. D. 1978. Design techniques for preventing input filter oscillations in switched-mode regulators. In *Proceedings of Powercon 5*, A3.1–A3.16.

Index

D

J

L

M

N

T

U

V

W

Z

Printed in the United States
by Baker & Taylor Publisher Services